TIDAL HYDRAULIC ENGINEERING

TIDAL HYDRAULIC ENGINEERING

S.N. Ghosh

Department of Civil Engineering
Indian Institute of Technology
Kharagpur

A.A. BALKEMA/ROTTERDAM/BROOKFIELD/1999

A.A. Balkema, P. O. Box 1675, 3000 BR Rotterdam, Netherlands
Fax: +31.10.4135947; E-mail: balkema@ balkema. nl;
Internet site: http://www.balkema.nl

Distributed in USA and Canada by
A.A. Balkema Publishers, Old Post Road, Brookfield, VT 05036-9704, USA
Fax: 802.276.3837; E-mail; Info@ashgate.com

ISBN 90 5410 735 9

PREFACE

The primary aim of this book is to provide students of civil engineering in universities with some background information pertaining to the planning, design and construction of various types engineering works involved in navigation, flood control, drainage and aquaculture in tidal environment.

Chapters 1 and 2 present the basic characteristics of astronomical tides, measurement of tides, current, discharges and processing of measured data. Subsequent chapters cover the following topics: (1) Fourier analysis and finite difference methods for solving tidal propagation problems; (2) sediment transport in a tidal environment, estimation of suspended load and morphological changes; (3) physical modelling of tidal flow, hydraulic aspects for the design of river training works, bank instability and erosion control measures; (4) engineering aspects of dredging, including radioactive tracer method for investigation of suitable dumping grounds and disposal of dredge spoils; (5) estimation of hydropower potential from tidal fluctuations; and (6) design of tidal outfall sluice gates and tidal channels for drainage of coastal areas and the design of sluice gates for water intake and drainage of aquaculture farms.

In the preparation of the manuscript the author has referred to lectures and course materials by eminent hydraulic engineers particularly by van Rijn of Delft Hydraulics Laboratory, demas (Dredging Engineering and Management Studies) Hague, Netherlands, Central Water and Power Research Station, Khadakwasla, Pune etc. Further papers published in various journals such Terra et Aqua, P.I.A.N.C, Central Board of Irrigation and Power, various national and international symposiums, conferences, publications of Irrigation Directorate of Government of West Bengal, Calcutta Port Trust including those furnished in the references. To all these organisations and individuals the author acknowledges his indebtednes and gratitude.

Although the selection of materials has been largely mine I am extremely grateful to the following persons for their invaluable assistance: Dr. S.S.

Chatterjee, Head Civil Engineering, Jalpaiguri Government Engineering College. Jalpaiguri, West Bengal, particularly for the chapter on Fourier Analysis; Dr. G.S. Reddy, Senior Research Engineer, Transoft International, Bangalore, particularly on chapter related to tidal computation and sediment transport. Next I am indebted to Dr. A.C. Ray, I.A.S., past Chairman of Calcutta Port Trust and presently Director, Institute of Social Welfare and Business Management, Calcutta; Sri H.P. Ray, I.A.S. currently Chairman Calcutta Port Trust. I am further grateful to Sri T. Sanyal, presently the Chief Hydraulic Engineer; Dr. A. Chatterjee, Dy. Director; Sri B. Ghatak, Assistant Scientific Officer and other colleagues of Calcutta Port Trust and to Dr. N. Dhang and Dr. Nageshkumar of the Civil Engineering Department at IIT, Kharagpur.

Finally, my heartful thanks to my wife Sm. Dyuti Ghosh, my daughter Pritha Ghosh and my son Archisman Ghosh.

Nov., 30th, 1997   (SOME NATH GHOSH)

IIT, Kharagpur   S.N. Ghosh

CONTENTS

CHAPTER 1

BASIC CHARACTERISTICS OF TIDES AND TIDAL PROPAGATION

1.1 ASTRONOMICAL TIDES

The basic theory regarding astronomical tides has been dealt with in books written by Defant [1], Neumann and Pierson [2] and Macmillan [3] and other specialised texts. Here emphasis is given to those aspects of tides which are of interest to engineers dealing with tidal and coastal engineering problems.

The gravitational attraction of the moon and sun on the earth and the equal and opposite centrifugal forces are the main forces responsible for the generation of tides. The sun's mass is approximately 2.7×10^7 times that of the moon, but it is far removed from the earth. The moon, in spite of its lesser mass exerts twice the effect of the Sun's gravitational force on the oceans, because it is closer to the earth.

The occurrence of tides can be explained from the law of gravitational attraction, i.e., the gravitational force varies inversely as the square of the distance between the two objects. The two bodies revolve about a common axis located about 2900 miles (or 4640 km) from the centre of the earth (Fig. 1.1) They are kept in orbit by gravitational pull which just equals the centrifugal force due to their rotation. Due to the large size of the earth the forces vary appreciably along its diameter. At the zenith the moon's gravitational pull exceeds the centrifugal force and as a result water tends to move towards the moon. At the nadir exactly the opposite happens, the water tends to move towards the point farthest from the moon. The result is two bulges or two high and two low tides each day as the earth rotates.

The earth rotates on it axis once in a 24-hour period. The moon, on the other hand, makes 1/29.5 of a revolution as the lunar orbit is 29.5 days. Figure 1.2 shows the orientation of the sun, moon and earth on the quarter points of the moon's revolution about the earth. The system is approximate and with reference to the sun. At the first and the third position, i.e., new moon and full moon, the solar and lunar forces reinforce and as such the

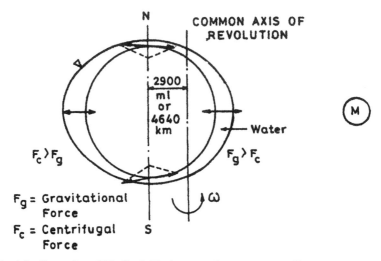

Fig. 1.1 Generation of idealised tide due to earth-moon system. F_g – gravitational force; F_c – centrifugal force

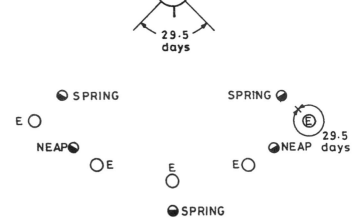

Fig. 1.2 Relative position of earth-moon-sun in a lunar month

highest or spring tide occurs, while at the second and fourth quarters the lowest or neap tides occur. Since each quarter is of 7.4 days duration the spring and neap tides are 14.8 days or roughly two weeks apart.

Due to the inclination of the plane of rotation of the moon around the earth there are times when the sun and moon are more in line with the earth than at other times. Normally this happens twice a year, near the spring and autumn equinoxes, although the timings vary year to year. During this time the greatest tidal range occurs during spring tides and the least tidal activity during adjacent spring neap tides. The whole propagation of spring neap tides and equinotical variations occurs in approximately 19 years.

The principal lunar and solar components are just two of over 360 active tidal components. They have periods ranging from about 8h to 18.6 yr. Each component has a period which has been determined from astronomical analysis. The amplitude and phase angle, on the other hand, depend on local conditions.

Doodson [4] provided a list of eight of the major components with their common symbols, period and relative strength, which are shown in Table 1.1.

Table 1.1   Major tidal components

Nature of tides	Description	Symbol	Period (solar h)	Amplitude (relative ratio strength)
Semi-diurnal	Principal lunar	M_2	12.42	100
	Principal solar	S_2	12.00	46.6
	Lunar component due to monthly variation in moon's distance from earth	N_2	12.66	19.1
	Solar-lunar constituent due to changes in declination of sun and moon throughout their orbital cycle.	K_2	11.97	12.7
Diurnal	Solar-lunar component	K_1	23.93	58.4
	Main lunar diurnal component	O_1	25.82	41.5
	Main solar diurnal component	P_1	24.07	19.3

Table 1.1 accounts for about 83% of the total tide generating force. In the Table N_2 is the tidal constituent due to the ellipticity of the moon's orbit, K_2 is associated with the variation of declination of both the moon and sun, K_1 is the most pronounced of the diurnal tides and is due to variation of both declinations, O_1 is lunar while P_1 is solar in origin.

The character of tide is determined in the following way. The ratio $F = (K_1 + O_1)/(M_2 + S_2)$ in which the symbols stand for the amplitudes of the concerned tidal constituents is computed. If the ratio is less than 0.25 the tide is characterised as semi-diurnal. The tide is considered mixed but predominantly semi-diurnal if the ratio lies between 0.25 and 1.5. The tide is said to be mixed but predominantly diurnal when the ratio lies between 1.5 and 3.0. If the ratio exceeds 3.0 the tide is diurnal. Figure 1.3 shows the common types of tides.

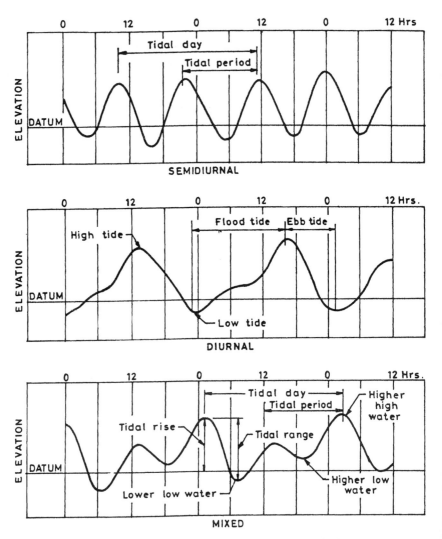

Fig. 1.3 Definition sketches of common type of tides

1.2 TIDAL DATUMS AND TIDAL RANGES

The oceans in the world consist of a network of interconnected basins. They cover approximately 70% of the global surface and have an average depth of about 4 km. The maximum horizontal dimensions may be as much as 10,000 km. The ocean boundaries are generally known as coastal areas and can be subdivided into beaches, shores, continental slopes and deep sea-floor.

The elevations of water in the coastal areas are expressed with reference to a variety of tidal datums in various parts of the world. Some of these datums and their reference levels are depicted in Fig. 1.4. Important terms with reference to tides are defined below.

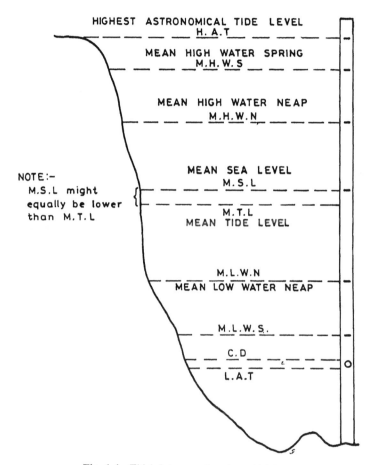

Fig. 1.4   Tidal datums and various tidal levels

1. Mean Sea Level (M.S.L.): the average height of the surface of the sea in all states of oscillations. This is taken as equivalent to the level which would have existed in the absence of tidal forces.
2. Mean Low Water Level (M.L.W.): the average of all the low water levels, i.e., low tide level.
3. Mean High Water (M.H.W.): the average of all the high water levels.
4. Mean Lower Low Water (M.L.L.W.): the average of only the alternate lower of low water levels (see mixed tide Fig. 1.3). M.H.H.W is the average of the alternate higher high tide levels.

5. Mean tide level; the level halfway between M.L.W. and M.H.W.

6. Range: The difference in level between consecutive high and low water.

M.S.L., M.L.W. and M.L.L.W are usually determined from tidal records covering a period of 19 yr. Gulf coasts are usually referenced to M.L.W. and the Pacific coast with respect to M.L.L.W. Elevations of land surface are usually referenced to M.S.L. It is therefore necessary to take appropriate care while combining topographical and hydrographic data.

The tidal range varies considerably. In the Baltic and Mediterranean seas tides are small. In Madras, the spring tidal range is 1.0 m. In India, the highest tidal range, above 11.0 m, occurs at Bhavnagar, Gujarat state. The world's highest tidal range is of the order of 30.0 m and occurs at the head of the Bay of Funday in New Brunswick. Mean and spring tidal ranges for some of the major ports of the world are given in Table 1.2.

1.3   TIDES IN COASTAL AREAS

The rise and fall of water in the oceans is propagated as a shallow-water wave, since the wavelength is large compared to the depth of water. The great primary wave of the Southern ocean has a maximum amplitude of 0.6 m, which is directly due to the influence of the sun and moon. Tidal waves in the Atlantic and Pacific oceans are generated from a primary wave and their amplitude is also 0.6.

As ocean tides approach a coast or an estuary they produce a coastal or estuarine tide.The behaviour of the tide is conditioned by the depth and shape of the ocean basin as well as the fact that motion takes place on a rotating spheroid. If the natural period of oscillation of water in a basin is close to one of the components of the tide-producing forces, oscillation of that period will build up. In many parts of the world the semi-diurnal tide becomes predominant. In others the diurnal components are amplified so much that the diurnal tide becomes dominant, even though the diurnal components of tidal forces are considerably less than the semi-diurnal components. Further, the effect of the earth's rotation on tidal propagation is fundamental. The tidal wave, instead of advancing in a direction determined simply by the east-west movement of the tide-generating force, rotates around a series of points of zero amplitude known as amphidromic points (Fig. 1.5). Thus the amplitude of tidal waves near a coast or estuary is substantially altered due to local effects.

It is not possible to calculate these local effects from the tide-generating forces. Instead, local tidal records can be analysed into components having the same periods as the harmonic constituents of the tide-generating force.

Table 1.2 Tidal ranges in some major ports.

Name of port	Mean range (m)	Spring range (m)
Antwerp (Belgium)	8.15	9.0
Auckland (New Zealand)	4.8	5.4
Bilboa (Spain)	2.74	3.6
Bombay (India)	2.66	3.6
Boston (USA)	2.9	3.35
Buenos Aires (Argentina)	0.67	0.73
Bay of Fundy (Nova Scotia)	12.7	14.5
Capetown (South Africa)	1.16	1.58
Cherbourg (France)	3.96	5.5
Dakar (Africa)	1.0	1.34
Dover (UK)	4.43	5.7
Genoa (Italy)	0.18	0.24
Gibraltar (Spain)	0.7	0.94
Hamburg (Germany)	2.32	2.44
Havana (Cuba)	0.31	3.64
Hong Kong (China)	0.94	1.62
Honolulu (Hawaii, USA)	3.64	0.58
La Guaira (Venezuela)	—	3.05
Lisbon (Portugal)	2.56	3.32
Liverpool (England)	6.45	8.24
Marseille (France)	0.12	0.18
Melbourne (Australia)	0.52	0.58
Mumbai (India)	2.66	3.6
Murmansk (USSR)	2.41	3.02
New York (USA)	1.34	1.62
Osaka (Japan)	0.76	1.0
Oslo (Norway)	0.31	0.34
Quebec (Canada)	4.2	4.72
Rangoon (Burma)	4.1	5.2
Reikjavik (Iceland)	2.8	3.8
Rio de Janeiro (Brazil)	0.76	1.07
Rotterdam (Netherlands)	1.52	1.65
San Francisco (USA)	1.21	1.74
Shanghai (China)	2.2	2.7
Singapore	1.7	2.21
Southampton (UK)	3.05	4.16
Sydney (Australia)	1.1	1.37
Vladivostok (USSR)	0.18	0.21
Yokohama (Japan)	1.07	1.43
Zanzibar (Africa)	2.66	3.8

Estimation of the amplitudes of these components and their phase relations enable tidal predictions to be made on the assumptions that each component will vary with time in the same way as the corresponding component of the

8

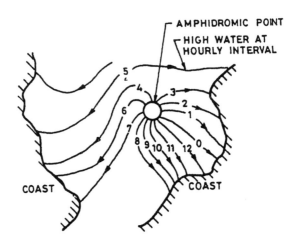

Fig. 1.5 Location of amphidromic point

tide-generating forces. Tidal predictions based on such harmonic analysis are published for all major and minor ports. In India they are prepared by the office of the Geodetic and Research branch of the Surveyor General of India, Dehra Dun, every year for the respective ports [5].

Due to the rotation of the earth, the Coriolis force acts, which is zero at the equator but appreciable at latitudes between 50° and 60° acting at right angles to the direction of flow. In the open sea this causes the current to rotate clockwise in the Northern Hemisphere and anti-clockwise in the Southern Hemisphere. That is, the current flows continuously with the direction changing cyclically through all points of the compass during a tidal cycle. When the tidal motion is constricted at entrance into estuaries, coastal inlets and harbours, reverse flows with larger velocities occur. The reversible type currents exhibit periodic changes in strength and direction corresponding closely to the periodic tidal range. Before reversing direction these currents pass through zero velocity or a slack condition. From the slacks the velocity increases gradually to a maximum, then decreases to zero, i.e., the current is reversed; the velocity then passes through a maximum in the opposite direction, again decreases to zero, thus completing a full cycle. Another type of tidal current, termed an hydraulic current, also reversible, occurs in canals connecting two independent tidal channels and inlets which connect the ocean with inland bodies of water. Such currents are caused not by tidal action directly, but by the phase difference of the tides in the two channels creating a head across the connecting canal.

Besides tidal currents, other types of non-tidal currents may also be present in coastal regions, namely:

(1) Wind-driven currents: Produced as a result of wind acting on the water surface. The velocity of the current generated by wind is about 2% of the wind velocity on the Pacific coasts and 0.5% on the Atlantic coasts.

(2) Freshwater currents: Caused by freshwater run-off from the watershed drained by a tidal estuary. The resultant direction is always towards the sea.

(3) Ocean currents: Permanent ocean currents comprise the mean oceanic circulation resulting from the difference in density caused by temperature difference.

(4) Littoral currents: In the ocean these flow parallel to the shore close to the coast and result from the action of waves on the beach. Waves strike the coast at a certain angle and the wave component is thereby turned parallel to the coast, which causes the littoral current.

Tidal and non-tidal currents often occur together and a current at any point is the result of these two classes of currents. It is sometimes necessary to isolate rotary currents and non-tidal currents when they occur simultaneously. The permanent set of currents at a station due to non-tidal causes may be obtained by the method of plotting the tidal current graph. Velocities and direction of the current averaged for a particular period are plotted as vectors radiating from a common origin. The hodograph thus obtained is generally elliptical. The direction and distance from the origin to the centre of gravity of the curve represent the permanent set of currents. The permanent set of currents may be due to the causes mentioned earlier.

1.4 PREDICTION OF TIDES

1.4.1 Harmonic Approach

A long record of the tidal curves measured at a site can be analysed for various lunar and solar constituents. Once these are known, the predictions of water levels can be done. In case the measurements are affected due to variations in weather, these variations have to be removed during analysis so that the constants obtained are related only to tide. The methods of analysis are harmonic or non-harmonic. In the case of harmonic analysis the results are affected by the limitations of the length of record available. Ideally, the record should be 19-year long but records or observations of a complete year are essential for a fully comprehensive analysis. This is because of the complex nature of the tide curves; a large number of higher harmonics is required to describe it accurately. In some cases, when the distortion of tide is small, the constants can be obtained from one-month data. The expression for instantaneous tidal elevation h above a chosen datum can be written as:

$$h = a_0 + \sum_{i=1}^{N} a_i \cdot \cos\left[2\pi t / T_i + \delta_i\right] \qquad \text{.... (1.1)}$$

where a_0 = vertical distance between the datum and M.S.L.; a_i = amplitude; T_1 = period and δ_i = phase angle of a particular component, i.e. (M_2, K_3); t = time and N = number of components used. The record is analysed to evaluate a_i and δ_i for each component of known period T_i.

Consider a tidal record from which we are to determine the amplitude a_i the phase angle δ_i, each of the seven important constituents: M_2, S_2, N_2, K_2, K_1, O_1 and P_1. The periods of each are 12.42, 12.00, 12.66, 11.97, 23.93, 25.82, and 24.07, h.

Let the tidal record be

$$\eta_t = a_0 + \sum_{i=1}^{N} \left[a_i \sin\left(2\pi t / T\right) \cos \delta_i + a_i \cos\left(2\pi t / T_i\right) \sin \delta_i\right] \quad \text{.... (1.2)}$$

Then consider for instance M_2 constituents; for this case T_i = 12.42 h.

Then
$$\sum_{i=1}^{N} \eta_i \, \sin \omega t = a_i / 2 \cos \delta_i \qquad \text{.... (1.3)}$$

$$\sum_{i=1}^{N} \eta_i \, \cos \omega t = a_i / 2 \sin \delta_i \qquad \text{.... (1.4)}$$

Divide T into 12 equal intervals. Therefore each interval is $T/12$, $n = 13$, $t_1 = 0$, $t_2 = T/12$, $t_3 = 2T/12$ etc. For each value of t_1, $t_2 \ldots t_n$, the values of η_1, $\eta_2 \ldots$ are obtained from tidal records. Also calculate the product of $\eta_i \sin w t_i$ etc. Finally, obtain the sum and obtain eqns. (1.3) and (1.4). From these obtain a_i and δ_i, i.e., the amplitude and the phase of the N_2 constituents. proceeding exactly in a similar way, find out the tidal constituents for S_2, N_2, etc. Then the prediction of resultant tide from constituent characteristics can be done on a computer.

1.4.2 Non-harmonic Approach

In the case of non-harmonic analysis it is necessary to compare the observations at a standard port. Such ports normally have long-period data for which detailed analysis has already been done. Comparisons are carried out graphically and provide the relative levels and time differences, thereby describing the tidal movements in the estuary with reasonable accuracy. On plotting the observations of high water at the standard port against those for the same event, at each site in the estuary a correlation can be made. Similar plots can be done for low water, low and high water times. The differences in time are plotted against the time of high and low water at the standard port. A curve is then drawn which is approximately sinusoidal in nature. The height plot will then indicate the locations where the low water levels are not the actual ones but dependent on river flow. It is possible to have a correlation between the volume of river flow and the low water levels recorded during the survey. The observed low water can then be corrected to the level expected for a certain river flow. Normally the standard adopted is a dry weather flow and an indication is given of the amount of rise to be expected in the rainy season. High-water effects are seldom noticed in an estuary. However, one notices such changes in these levels in the rivers themselves.

1.5 MEASUREMENT OF TIDE

1.5.1 Objective

The measurement of tide is very important for practical engineering purposes as well as for tide of prediction. The most important purpose is construction navigational charts, for which echo-sounding surveys have to be based on a datum. In coastal areas the reference level, commonly termed as the sounding datum, is usually fixed at the level of the lowest predicted tide. Sometimes the reference level or the sounding datum is a chart datum fixed at a certain tide level. The sounding datum may be a totally arbitrary. level, but is always expressed as a certain distance below a permanent mark, such as a bench-mark of the land levelling system or the sill of a dock. Once this has been done tide measurements will need to be taken during the surveys. Secondly, to provide predictions for navigational purposes it is necessary to have continuous records, as many estuaries show considerable changes in water level with atmospheric pressure and wind. For the navigation of vessels with minimum underkeel clearance it is necessary to provide actual levels rather than predicted levels. Thirdly, for prediction of tides tidal records for a period of one year are required, although ideally it should be for nineteen years. This will provide the astronomical constants for prediction.

1.5.2 Tidal Gauges and Their Location

Water levels are obtained through gauges. The gauges are either visually operated or function through mechanical or electrical instrumentation. The data is recorded in the form of a graph.

12

To locate the gauges the following factors have to be taken into account:

1. Location or position shall not be unnecessarily exposed to winds; special devices may be required to reduce the effects of waves and turbulence.
2. Accessibility of the gauge even in high-water conditions needs to be assured and the reach of the gauge should be sufficient to cover the total variation of the water level.
3. Zero level of the gauge should be connected to a reference level.

For regular monitoring of the proper functioning of the gauge, a stable bench-mark should be established in the vicinity of the gauge. Non-recording gauges which are visually read are relatively cheap and easy to install. They require skilled observers, however, even then the possibility of human error has to be taken into account.

Staff gauges (Fig. 1.6): These are the simplest type gauges, consisting of a graduated gauge plate fixed vertically to a stable structure such as the wall or pier of a bridge. The gauge can be read directly at any moment.

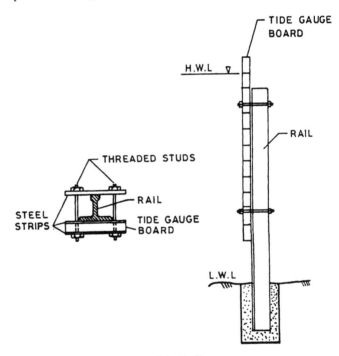

Fig. 1.6 Staff gauge

Float gauges (Fig. 1.7): Simply constructed from a float moving in an open pipe along with a graduated steel tape, counterweight and a pulley. The reading can be made directly through a pointer moving along the tape.

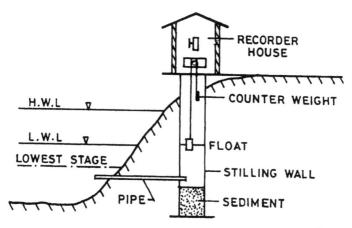

Fig. 1.7 Float gauge

Pneumatic gauges (Fig. 1.8) can be of two different operating systems. One is the diaphragm type and the other is the bubble type. The basic principles of both the systems is that the head of water above the diaphragm or air outlet is measured through a pressure gauge. These readings are then converted into the water level.

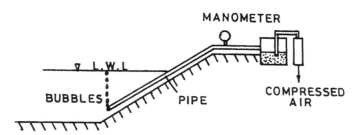

Fig. 1.8 Pneumatic gauge

For automatic recording of the continuous movement of the water level one can adopt either the analogue system or the digital system. Also, data can be telemetered to a central station. The analogue system provides a graphic record proportional to the actual rise and fall of the water level with respect to time. The graph is recorded on graph paper affixed to a rotating drum and actuated by a clock mechanism. The digital recorder punches the code values of a stage on paper tape at preselected time intervals. Telemetry systems are used to provide water-level data from remote stations, either as a continuous indication of stage or as instantaneous information on call or at predetermined intervals. The receiving original signals from the electronic tide gauge installations are then converted to digital output or readout.

1.5.3 Calibration of Tidal Gauges

The tidal gauge needs to be calibrated to ensure that the heights are correctly recorded or else the corrections to be applied must be known. After installation of the gauge and the usual tidal staff a series of comparisons between the two should be recorded over a complete tidal cycle, preferably the highest spring tide, to cover the full range of the tidal gauge. On being plotted the visual values and gauge values should form a straight line on an ordinary graph sheet, the slope of which is dependent on both the instrument in use and the site at which it is installed. In case the recorder is changed or moved to a new site a new calibration will be required. As the observations progress it is possible to add daily checks to the diagrams and if they are made at about the same time each day they will be spaced out over the calibration line. Thus it is possible to check the initial calibration and also indicate when any change in zero level takes place, which may be due to loss of air pressure or movement of the diaphragm box. after repairing the defect a new calibration has to be carried out.

1.6   TIDES IN ESTUARIES

Most rivers enter the sea where there is enough tidal rise and fall to modify the flow near their mouths. The part of the river system in which the river widens under the influence of tidal action is called an estuary. Cameron and Pritchard [6] defined estuary as a 'semi-enclosed body of water, one end of which is subject to tidal variation and the other end of which is subject to freshwater flow'.

The rise and fall of the water surface at the entrance of an estuary causes surface gradients. This results in the propagation of a gravity wave into the estuary. The propagation rate mainly depends on the depth of water, in other words, on the tidal range at the mouth. The speed of propagation of waves is different from that of the fluid through which they move. The wave travels with a celerity $= (gh)^{1/2}$, relative to the water. It has a value of about 10 m/s in 10 m depth. Another factor of considerable influence on the penetration of tide is the discharge of the river (Fig. 1.9).

In most estuaries the depth of water is not high, especially at low water. This causes considerable friction between the water and the bed, which is most noticeable during the falling tide. This effect can be seen on tide gauge records where the slope of the rising tide will be very much steeper than that of the falling tide. Figure 1.10 shows the typical tidal curves for three positions in a tidal river. Further, the time of rise from low to high water decreases at points farther inland and will be much smaller than that of the fall. Shallow-water effects cause the crest of the tidal wave to travel up the estuary much faster than the trough. In some cases the effect of

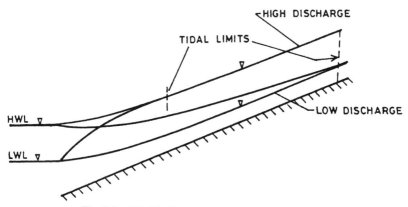

Fig. 1.9   Tidal limit as a function of upland discharge

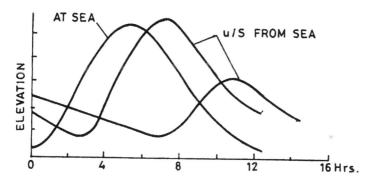

Fig. 1.10   Typical tidal curves at two locations upstream from sea

friction combined with the slope of the bed reaches a critical state at which bores are formed. Under these circumstances the onset of the crest of the tide takes the form of a breaking wave which can be clearly seen advancing up the estuary. Besides from the breaking effect, a very rapid reversal of tidal flow takes place as the wave pases which, coupled with rapid rise in the water level, can be very dangerous to shipping and persons using the estuary.

During ebb tide surface gradients are set up at the seaward end which are initially similar in magnitude to those occurring during flood tide but in the opposite direction. As the water level falls, the rate of wave propagation decreases and the influence of the rate of fall of the sea becomes weaker. The ebb flow then becomes a process of drainage under gravity.

1.6.1   Behaviour of an Estuary

The behaviour of an estuary is discernible from the record of the rise and fall of the water surface at several stations along its length. The Hooghly

River has a moderate tidal range of about 5.5 m and tidal length of about 300 km. In this river when high water occurs at the mouth, the previous high water has just reached the tidal limit.The other important effect of tidal propagation is the M.T.L. The M.T.L. at the sea face at Saugor is the same for spring and neap tides. At stations farther inland the M.T.L. during springs is higher than neaps, which means that during the spring tide cycle a large volume of water can accumulate in the upper part of an estuary, which causes an increase in salinity during spring tides and a decrease during neaps. It also contributes to the net landward movement of sediment during spring tides and net seaward movement during neaps with peak of sediment concentration occurring later than the highest spring tides.

Normally both the beds and banks of the estuaries are erodable and in such cases the estuary widens exponentially towards the sea end. The rate of sediment transport per unit width is usually greatest in the middle reaches of an estuary; contrarily the total rate of sediment transport may be greatest near the sea since the widths are very large. The largest rate of sediment transport in the absence of river flow is landward and the total quantities in movement are greatest near the mouth, although the greatest rates of transport per unit width occur in the middle reaches. The overall trend of tidal action in the absence of river flow is to transport sediment landwards.

The equilibrium of an estuary can be maintained if the quantities of solids, fresh water and minerals in solution each remain in balance. This requirement of continuity of matter has a far reaching effect on the behaviour of the estuary. Fresh water entering the estuary must leave at the same rate, averaged over a period of several weeks, if the system is in equilibrium. Water leaving an estuary mixes with saline water until it is indistinguishable from it. The process is a gradual one and detected many kilometres offshore. The movement of fresh water from the estuary mouth to the sea is accompanied by the movement of saline water entrained with it (Fig. 1.11). This saline water needs to be replaced if equilibrium is to be maintained. It is therefore necessary to replace the exact amount of salt water removed after entraining with freshwater per unit time by an equal amount of influx of water with dissolved salts. This takes place towards the region near the bed. The density difference between the water at the seaward end and water entering from rivers causes landward movement of water near the bed and a compensating seaward movement near the surface. Fine sediments can be carried landwards in suspension due to this up to the landward limit of density gradients. The position varies depending on river discharge.

When turbulent mixing is intense there is only a small difference in density over the depth. At any point there must be a horizontal gradient of

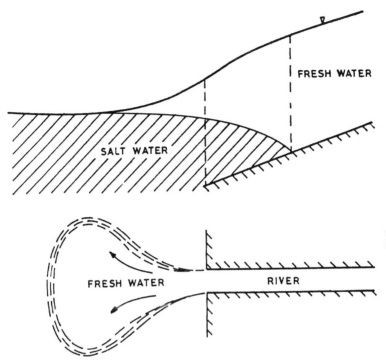

Fig. 1.11 Penetration of salt water

density ranging from 1000 kg/m^3 at the upper tidal limit to 1026 kg/m^3, i.e., the density of sea water at some offshore distance. The horizontal force due to such a density gradient increases with depth below the surface and is always directed in the direction of decreasing density. This gives rise to a small landward force that is zero at the surface and reaches maximum at the bed.

At low flow rates mixing between fresh river and sea water occurs at the narrow interface between the layers. Some sort of mixing will occur at the interface and to maintain the equilibrium landward movement of saline water in the lower layers will result. This can be explained as follows (see Fig. 1.12).

If there is a fresh water flow Q_f into the channel and as result there is a landward flow of Q_s in any part of the cross-section, there must be seaward flow of $Q_f + Q_s$ in the remainder of the cross section. Similarly for salt the net flow for equilibrium at any section must be zero.

Stratified flow occurs in estuaries having weak tidal action, i.e., in estuaries with a small range or steeply sloping beds, which means small tidal storage volume. It can occur during neap tide in estuaries that are partially mixed. Conversely, well-mixed estuaries have high tidal activities.

18

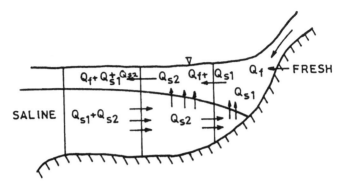

Fig. 1.12   Effect of entrainment of salt and fresh water

Based on degree of stratification, attempts have been made to classify estuaries. According to Harleman [7] the numerical value of the estuary number $(P_t F_0^2 / Q_f T)$ determines the degree of stratification.

A large value indicates well-mixed estuaries where P_t is the volume of the tidal prism, defined as the volume entering the mouth of the estuary during the rising tide; Q_f rate of inflow of fresh water. T the tidal period. F_0, Froude number $U/(gh)^{1/2}$ where U is the maximum flood-tide velocity average across the mouth; and h the depth below mean water level at the mouth of the estuary.

LITERATURE CITED

1. Defant, A. (1961). *Physical Oceanography*, Vol. 2. Pergamon Press, Oxford, U.K.
2. Neuman, G. and Pierson, W.J. (1966). *Principles of Physical Oceanography*. Prentice-Hall, Englewood Cliffs, New Jersey.
3   Macmillan, D.H. (1966). Tides, C.R. Book.
4. Doodson, A.T and Warburg, H.D. (1941). *Admiralty Manual of Tides*. HMSO.
5. Hooghly River Tide Tables. Office of Surveyor General of India, Dehra Dun (yearly).
6. Cameron, W.M. and Pritchard, D.W. (1963). *Estuaries in the Sea*, vol. 2 (M.N. Hill, ed.). John Wiley and Sons, New York.
7. Harleman, D.R.F. and Abraham, G. (1966). One-dimensional analysis of salinity intrusion in the Rotterdam Waterway. *Delft Hyd. Lab. Pub*. no. 44.

CHAPTER 2

TIDAL MEASUREMENT AND ANALYSIS

2.1 INTRODUCTION

The measurement of tidal flow and other parameters in a tidal river is usually carried out based on field observations. The measurements include some form of current velocity, water temperature, salinity and sediment. The sediment measurement and its computational aspect have been dealt with in a separate chapter. The representativeness of any set of measurements is dependent on the choice of sampling locations, sampling rate and duration of study. A rational procedure of analysis and synthesis of data is also an essential requirement.

Sampling design: It is normally advisable to choose carefully a suitable sampling design. Location of stations and the number of vertical measurements per station should be selected with great care. For the measurement of discharge it is usually necessary to include a minimum of 3 stations in a cross-section. In the vertical it is usually necessary to make measurements at least at five locations between the surface and the bottom if the water depth permits to ensure that the profiles are adequately defined.

2.2 MEASUREMENT OF TEMPERATURE AND SALINITY

Temperature can be measured in many ways. Nowadays thermistors have replaced ordinary thermometers as preferred sensors for temperature measurements. A thermistor can be made very small, responds almost instantaneously to temperature variations and is capable of measuring temperature with a precision of 0.001°C. Thermistors are commonly integrated into salinity-temperature-depth sensors used for on-site measurements.

The salinometer is now the most frequently used instrument for salinity determination. This instrument measures water conductivity and temperature from which salinity can be computed. The electrical conductivity of water is the inverse of its resistance and is expressed in units of ohm^{-1} cm^{-1} or mho cm^{-1}. The relation between salinity and electrical conductivity/temperature is not a simple one [1].

2.3 VELOCITY MEASUREMENTS

The surface velocity of flow is normally measured using a float. For this purpose two stations are fixed one at the upstream and the other downstream of the control section at a known distance L. To measure surface velocity during flood tide one surface float with a post and flag is introduced at the midpoint òf the river cross-section at station A, i.e., downstream station. The time taken by the float to traverse the distance L between A and B is noted. The surface velocity v_s is obtained as L/t. During ebb tide the float is introduced at upstream station B. To obtain the velocity at different times floats are introduced at different times depending on the desired time interval. The flow direction is ascertained by observing the direction of movement of floats. The average flow velocity v_a along the depth can be obtained as $v_a = (5/6)v_s$.

The current velocity varies greatly both in magnitude and direction relative to its resultant value over the predominant tidal period. The current velocity is obtained by a currentmeter which can either be suspended from the surface or installed at a fixed distance above the bottom. The direction is usually sensed through a vane which is allowed to rotate in the horizontal plane or the meter itself swings with the current.

2.4 DISCHARGE MEASUREMENTS

2.4.1 Cubature Method

In this approach the discharges are computed from the changes in the volume of water contained in a reach of a tidal channel during a given time interval. It is equal to the differences between the volume of water in the channel at short intervals of time at a number of stations along the channel up to tidal limit. The areas of water surface corresponding to various water levels are determined from each reach. Changes in the volume of water during the specified time intervals are computed. Freshwater discharges if any up to the channel are determined by a standard method and algebraically allowed for in the computation [2]. Figure 2.1 shows typical simultaneous tidal curves. With reference to Fig. 2.2 let A_1A_2 and B_1B_2 be the simultaneously observed water surface levels at the beginning and at the end of the time interval between sections (1 – 1) and (2 – 2) along the river. The water surface profile has risen and so represents flood in Fig. 2.2, $B_1 B_2$; ebbing is represented with water surface level lowering from B_1B_2 to A_1A_2. Upland discharge should be measured separately at a station

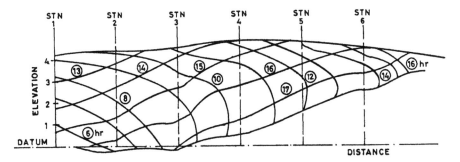

Fig. 2.1 Typical simultaneous tidal curves

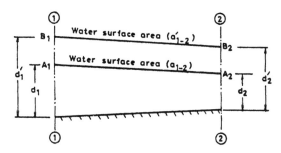

Fig. 2.2 Simultaneous water level over a time interval

beyond the reach of tidal influence. To compute the volume of water flowing seaward at a section the freshwater flow over the given time interval is added to the seaward flow computed from the change in volume of tidal prism during the same interval. The changes in volume according to the Pillsbury [3] method is:

$$V_{1-2} = [(a_{1-2} + a'_{1-2})]/2 \ [(d'_1 - d_1) + (d'_2 - d_1)]/2 \quad \dots (2.1)$$

where (a_{1-2}), (a'_{1-2}) are area of water surface between stations 1 and 2 at the beginning and end of time interval; $(d_1, d_2),(d'_1, d'_2)$ are depths of water levels at the beginning and end of time interval at stations 1 and 2 respectively; $(w_1, w_2),(w'_1, w'_2)$ are the corresponding river widths at the beginning and end of time interval at sections $(1 - 1)$ and $(2 - 2)$ respectively. L_{1-2} is the distance between sections, Q the upland discharge and t the time interval.

For the stations at 1, 2, 3... up to the tidal limit, the total volume flowing across section 1 in the time interval in question is

$$V = (V_{1-2} + V_{2-3} + \dots + \text{up to tidal limit}) - Qt. \quad \dots (2.2)$$

In computing V_{1-2} etc. the correct sign is used for flood (+ive) and for ebb (–ive). The tidal discharge is thus (V/t) and if there are no side storage areas in the entire length, the area of water surface between the two stations can be estimated by Kilford's [2] method, i.e.,

$$a_{1-2} = [w_1 + w_2] L_{1-2}/2, \quad a'_{1-2} = [w'_1 + w'_2] L_{1-2}/2. \qquad \ldots (2.3)$$

So $\quad V_{1-2} = L_{1-2}/8 \,[w_1 + w_2 + w'_1 + w'_2] \,[d'_1 + d'_2 - d_1 - d_2] \qquad \ldots (2.4)$

The equation is simpler and is suitable when the channel is prismoidal in shape.

2.4.2 Velocity Area Method

This method is based on the measurement of water surface level, current velocity and cross-section of the river at the station of measurement. (Fig. 2.3). The measurement of water surface level is done through a tide gauge installed at the station as discussed earlier. In the case of a staff gauge the height is recorded at intervals of 30 minutes throughout the tidal cycle. The cross-sectional shape of the river at the control station is determined by stream gauging. The cross-sectional areas of flow at different times A_1, A_2, A_3 at the station are determined by superimposing the water surface at the corresponding time on the cross-sectional shape of the river. The discharges Q_1, Q_2, Q_3, etc. at different time t_1, t_2, t_3, etc. are obtained as:

$$Q = A_1 \overline{V}_1, \; Q = A_2 \, \overline{V}_2, \; Q = A_3 \, \overline{V}_3 \text{ etc.} \qquad \ldots (2.5)$$

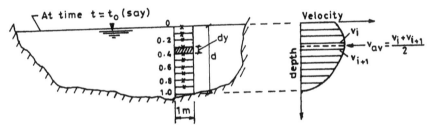

Fig. 2.3 Velocity measurement at relative depths 0 to 1

When velocity measurements at a number of points in a vertical are available and if there are m verticals in a cross section, the discharge can be estimated as follows. Let q be the discharge per unit cross-sectional area, i.e., q in m²/s. The total discharge through the cross-section can then be obtained, after integration of m estimates of q numerically:
Where q can be expressed as

$$q = 1/T \int_0^T \overline{V}(t) \, y(t) \, dt \qquad \ldots (2.6)$$

where $y(t)$ is the time varying water depth. $\overline{V}(t)$ is the time varying depth averaged velocity normal to the cross-section and is computed from:

$$\overline{V}(t) = 1/10[1/2\ V(\eta_0 \cdot t)] \int\limits_{j=1}^{9} V(\eta_j,\ t) + 1/2V(\eta_0,t) \quad \cdots (2.7)$$

where η_1, η_2...η_n represent non-dimensionalised depths between surface and bottom; t = sampling time interval.

The surface and bottom velocities are weighted by 1/2 as these boundary values are only representative over half the depths as compared to the velocity values at η_1 through η_9. Actually it is not necessary to include $V(\eta_{10},\ t)$ in the equation as the velocity of water sediment interface is always zero. It has been included to point out that for salinity, temperature, etc. the term is non-zero. The discharge is then estimated from:

$$q = 1/\eta \left[1/2\ \overline{V}(t_0)\,y(t_0) + \sum_{k=1}^{\eta-1} V(t_k)\,y(t_k) + 1/2\,\overline{V}(t_\eta)\,y(t_\eta) \right] \quad \cdots (2.8)$$

With m stations in a cross-section the total net cross-sectional discharge Q is given by:

$$Q = \sum_{i=1}^{m} w_i\,q_i \quad \cdots (2.9)$$

where w equals width over which q is assumed to be representative.

For each parameter computation is made of time averaged values for the non-dimensionalised depths = 0, 0.1, 1.0. Measurements should be made at a constant sampling interval dt over one complete tidal cycle. The net value of one of the measured parameters X at a given non-dimensionalised depth η is expressed as $X(\eta)$ and computed from

$$X(\eta_j) = \frac{1}{n}[1/2X(\eta_0,\ t_0) + \sum_{k=1}^{n-1} X(\eta_j\ t_k) + 1/2X(\eta_j,\ t_n)]$$

where $j = 0, 1, \ldots 10$, η_0, η_1, η_{10} represent non-dimensionalised depth between surface and bottom; t_0 is the starting time; t_n is the time at the end of the tidal cycle and

$$t_k = t_0 + kdt$$

for $\qquad k = 1, 2,\ n-1$

LITERATURE CITED

1. Kjerfve, B. (1975). Measurement and analysis of water current, temperature, salinity. Belle W. Baruch Institute of Marine Biology and Coastal Research. University South Carolina, USA (no. 172).
2. River Behaviour, Control and Training (1971). Chapter IX, Tidal Training for Navigation. CBI and Power, Govt. of India, New Delhi.
3. Pillsbury, B. (1956). Tidal Hydraulics. Corps of Engineers, Vicksburg, PA.

APPENDIX: NUMERICAL EXAMPLE

The following are the hourly tidal velocity measurements in a section.[1] Calculate the net discharge per unit width, depth averaged net velocity, percentage of ebb and flood flow. Give a graphic plot time vs velocity, elevation and discharge.

Relative depth					Velocities in cm/s at hourly intervals								
η	t_0	t_1	t_2	t_3	t_4	t_5	t_6	t_7	t_8	t_9	t_{10}	t_{11}	t_{12}
0	−2	−3	−9	−46	−60	−65	−7	73	100	89	55	24	10
0.1	−4	−6	−20	−50	−63	−65	−14	64	85	80	54	24	8
0.2	−6	−12	−33	−55	−67	−65	−29	60	83	78	63	24	2
0.3	−10	−18	−40	−55	−67	−66	−33	46	87	76	71	24	0
0.4	−12	−26	−48	−56	−66	−65	−28	40	87	73	71	22	8
0.5	−12	−29	−45	−60	−65	−64	−25	37	79	70	71	16	0
0.6	−20	−29	−40	−64	−62	−61	−22	37	73	69	71	10	0
0.7	−22	−17	−40	−60	−50	−66	−21	37	67	68	53	8	0
0.8	−20	−12	−38	−56	−50	−46	−20	36	59	68	42	5	0
0.9	−18	−8	−29	−44	−37	−24	−18	32	44	60	37	2	0
1.0	0.0	0.0	0.0	0.0	0.0	0.0	0.0	0.0	0.0	0.0	0.0	0.0	0.0
depth (m)	6.1	6.4	6.9	7.4	7.9	8.1	8.2	8.2	7.6	7.1	6.7	6.5	6.3

Solution

For computing net discharge per unit width it is necessary to compute depth averaged velocity for each time from eqn. (2.7). For $t = t_0$ it yields

$$\overline{V}(t_0) = 1/10 \{-1/2 \; 2 - 4 - 6 - 10 - 12 - 12 - 20 - 22 - 20 - 18\}$$
$$= -12.5 \text{ cm/s}$$

Similarly, depth averaged velocity is computed for t_1 through t_{12} and these are respectively, −16.0, −34.0, −52.0, −55.0, −55.5, −21.4, 42.5, 72.4, 68.7, 56.0, 14.7, 1.5 in cm/s. The next step is to calculate the discharge according to eqn. (2.8):

$$q = 1/13 \{-1/2 \times 125 \times 6.1 - .16 \times 6.4 - 34 \times 6.9 \\ - 52 \times 7.4 - 55 \times 7.9 - .555 \times 8.1 - .214 \times 8.2 + .425 \\ \times 8.2 + .724 \times 7.6 + .687 \times 7.2 + .56 \times 6.7 + .147 \\ \times 6.5 + .015 \times 6.3\} = .020 \text{ m}^3/\text{s/m}.$$

The graphic plot of depth, velocity and discharge with time is shown in Figure 2.4.

Check Calculation

Flood discharge. With reference to Figure 2.4 the flood discharge volume equals – 65877.75 m^3

Ebb discharge volume equals = 66,798 m^3

Net discharge volume equals = 65,877.75 + 66,798 = 920.25 m^3

Discharge rate = 920.25 /(12 × 60 × 60) = .0213 m^3/s/m

Percentage of flood = 65,877.75/(65,877.75 + 66,798) = 49.7 and percentage of ebb = 50.3.

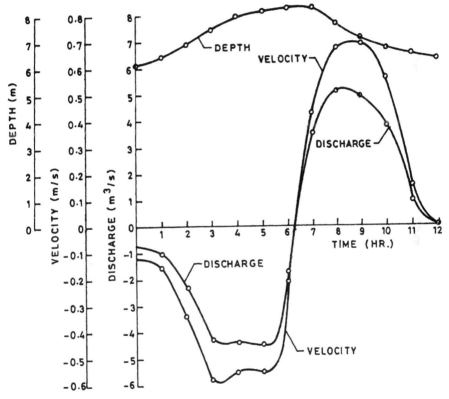

Fig. 2.4 Plot of variation of depth, velocity and discharge with time

CHAPTER 3

FOURIER ANALYSIS OF TIDAL FLOW PROBLEMS

3.1 BASIC ASSUMPTIONS

The basic assumptions made in the Fourier analysis of tidal flow problems are:

1. The tide is simple harmonic.
2. The coefficients of each term are constant between two consecutive sections of the channel.
3. At any distance from the mouth of the channel the cross-section is rectangular.
4. The channel bed is horizontal.
5. The channel is closed at the upstream end.

3.2 FOURIER SERIES: DEVELOPMENT OF TERMS

Let $h(0, t)$ represent the water level with respect to mean water level as a continuous known function of time at the mouth of the river, $x = 0$. Since the channel is assumed to be closed at the upstream end, there is no upland discharge,

$$h(0, t) = \sum_{n=1}^{\infty} h_n(0) \cos(n\omega t + \alpha_n(0)) \qquad \dots (3.1)$$

where $h_n(0)$ and $\alpha_n(0)$ are amplitude and phase respectively of the harmonic component h_n. The water level with respect to mean level and discharge along the river can be expressed as:

$$h(x, t) = \sum_{n=1}^{\infty} h_n(x) \cos(n\omega t + \alpha_n(x)) \qquad \dots (3.2)$$

$$q(x, t) = \sum_{n=1}^{\infty} q_n(x) \cos (n\omega t + \beta_n(x)) \qquad \dots (3.3)$$

Equations (3.2) and (3.3) in complex notations can be expressed as:

$$h(x, t) = \sum_{n=1}^{\infty} H_n(x) e^{in\omega t} + H_{-n}(x) e^{-in\omega t} \qquad \dots (3.4)$$

$$q(x, t) = \sum_{n=1}^{\infty} Q_n(x) e^{in\omega t} + Q_{-n}(x) e^{-in\omega t} \qquad \dots (3.5)$$

where $H_n(x)$, $Q_n(x)$ are the complex amplitude and discharge of the n-th harmonic Fourier coefficient and $H_{-n}(x)$ and $Q_{-n}(x)$ are the complex conjugate terms.

3.3 DEVELOPMENT OF EQUATIONS FOR $H_n(x)$ AND $Q_n(x)$

Equations for $H_n(x)$ and $Q_n(x)$
(n = 1,2,3...) can be derived by substituting (3.4) and (3.5) into basic differential equations for tidal motions. The basic equations are respectively,

$$\partial h/\partial x + (1/gA)\, \partial Q/\partial t - b/(gA)^2\, Q\, \partial h/\partial t + [1/(C^2RA^2)]Q\, |Q| = 0 \dots (3.6)$$

$$\partial Q/\partial x + b\, \partial h/\partial t = 0 \qquad \dots (3.7)$$

where, g = acceleration due to gravity;
R = hydraulic radius;
b = width of the channel;
A = cross-sectional area.

Equation (3.6) is non-linear because the coefficients of the equation depend on distance and time and also the resistance term is not linear in the independent variable Q. The equations therefore have to be transformed before functions (3.4) and (3.5) can be introduced. For this Fourier development of the coefficients is required. This has been furnished in the Appendix of this Chapter.

3.4 EQUATIONS OF MEAN TIDE

Equations of mean tide can be obtained from eqns. (3.6) and (3.7) after introducing the various Fourier developments of Q, $\partial Q/\partial x$, $\partial Q/\partial t$, h, $\partial h/\partial x$, $\partial h/\partial t$, b, $1/gA$ and $1/(C^2R A^2)$ where Q and h are defined by

$$\left.\begin{array}{l} Q = Q_1(x)\, e^{i\omega t}\, Q_{-1}(x)\, e^{-i\omega t} \\[2mm] h = H_1(x)\, e^{i\omega t} + H_{-1}(x)\, e^{-i\omega t} \end{array}\right\} \qquad \dots (3.8)$$

$$\partial Q/\partial x = d/dx \; Q_1(x) \; e^{i\omega t} + \text{c.c} \qquad\qquad \Bigg\}$$

$$\partial Q/\partial t = i\omega Q_1(x) \; e^{i\omega t} + \text{c.c} \qquad\qquad \quad \dots \; (3.9)$$

$$\partial h/\partial x = d/dx \; H_1(x) \; e^{i\omega t} + \text{c.c} \qquad\qquad \Bigg\}$$

$$\partial h/\partial t = i\omega H_1(x) \; e^{i\omega t} + \text{c.c} \qquad\qquad \quad \dots \; (3.10)$$

$$b\partial h/\partial t = b \left(i\omega H_1(x) \; e^{i\omega t} + \text{c.c} \right)$$

$$(1/gA) \; \partial Q/\partial t = \left[\left(i\omega \, m_0 \, Q_1 e^{i\omega t} - i\omega m_1 \; H_1 \; Q_{-1} \right) + \text{c.c} \right] \qquad \dots \; (3.11)$$

$$Q \, |Q| \, (C^2 A R^2) = \left[r_1 \, H_1 \, C_{-1} \, Q_{-1} + \left(r_0 \, C_1 \, Q_1 + r_2 H_1^2 \; Q_{-1} \, C_{-1} \right) e^{i\omega t} + \text{c.c} \right]$$
$$\dots \; (3.12)$$

$$(b/A^2) \; Q\partial h/\partial t = \left[\left(\omega b / A_0^2 \right) i \left(Q_{-1} \, H_1 \right) + \text{c.c} \right] \qquad \dots \; (3.13)$$

Introducing all the above developments in eqn. (3.6) and simplifying, we have

$$dH_1/dx + \left(16/3\pi \; r_0 \; |Q_1| + i\omega m_0 \right) Q_1 + 16/3\pi \; r_2 \; H_1^2 |Q_{-1}| Q_{-1} = 0 \quad \dots \; (3.14)$$

Similarly from the continuity equation (3.7), we have

$$dQ_1/dx + i\omega b H_1 = 0 \qquad\qquad \dots \; (3.15)$$

Equations (3.14) and (3.15) are for the complex amplitude and discharge for the main tide. The factors m_0, r_0, r_2 and z are furnished in the appendix.

3.5 SOLUTION OF EQUATIONS

Omitting the second order term containing the sine and cosine functions, eqn. (3.14) can be written as:

$$dH_1/dx + (i\omega m_0 + r_0^*) \; Q_1 = 0 \qquad\qquad \dots \; (3.16)$$

where $m_0 = (1/gA_0)(1 + 1/2 \; z)$;   and $r_0^* = 16/3\pi \; r_0 \; |Q_1|$

or $\qquad\qquad r_0^* = 16/3\pi \; C^{-2} A_0^{-2} \; h_0^{-1} \; (1 + 3z) \; |Q_1|$

Eliminating Q_1 from (3.16) with (3.15), one obtains.

$$d^2 H_1/dx^2 + \left(\omega^2 b \, m_0 - i\omega b \, r_0^* \right) H_1 = 0$$

Let $\qquad\qquad k^2 = - \left(\omega^2 b \, m_0 - i\omega b \, r_0^* \right)$, so we have

$$d^2H_1/dx^2 - k^2H_1 = 0 \qquad \text{.... (3.17)}$$

Let $\qquad H_1 = e^{mx}$ be the solution of the equation then,

$$H_1(x) = C_1\, e^{kx} + C_2\, e^{-kx}$$

$$= A\, \cosh kx + B\, \sinh kx \qquad \text{.... (3.18)}$$

where $A = (C_1 + C_2)$ and $B = (C_1 - C_2)$

Again from eqn. (3.15)

$$dQ_1/dx = -i\omega b\, (A\, \cosh kx + B\, \sinh kx) \qquad \text{.... (3.19)}$$

therefore $\qquad Q_1(x) = -i\omega b\, (A/k\, \sinh kx + B/k\, \cosh kx) \qquad \text{.... (3.20)}$

at $\quad x = 0,\ H_1(x) = H_1(0),\ Q_1(x)\, Q_1(0)$

So $\qquad A = H_1(0);\ \ (-i\omega b/k)\, B = Q_1(0);\ \ B = -K/(i\omega b)$
$Q_1(0) = (ki\omega b)Q_1(0)$

$$H_1(x) = H_1(0)\, \cosh kx + ki/\omega b\, Q_1(0)\, \sinh kx \qquad \text{.... (3.21)}$$

$$Q_1(x) = Q_1(0)\, \cosh kx - i\omega b/k\, H_1(0)\, \sinh kx \qquad \text{.... (3.22)}$$

$$k = \pm\, \omega\, \sqrt{b m_0}\, (\theta + i\theta_1)$$

$$\theta = \left[-1/2 + 1/2 \left(1 + S_1^2\right)^{1/2} \right]^{1/2}$$

$$\theta_1 = \left[1/2 + 1/2 \left(1 + S_1^2\right)^{1/2} \right]^{1/2}$$

$$S_1 = r_0^*/(\omega\, m_0)$$

Let $\qquad \theta_2 = \omega\sqrt{b m_0}\ 1/2\, S_1;\ \theta_3 = \omega\sqrt{b m_0}\, ;\ \ \text{So } k = \pm\, (\theta_1 + i\theta_3)$

Introducing the positive value of k in eqn. (3.21) and simplifying, one obtains:

$$H_1(x) = \left[H_1(0)\, \phi_1(x) - 1/\omega b\, Q_1(0)\, (\theta_2\, \phi_2(x) + \theta_3\, \phi_4(x)) \right]$$
$$+ i\left[H_1(0)\, \phi_3(x) + 1/\omega b\, Q_1(0)\, (\theta_2\, \phi_4(x) - \theta_3\, \phi_2(x)) \right] \qquad \text{.... (3.23)}$$

where $\qquad \phi_1(x) = \cosh \theta_2(x)\, \cos \theta_3(x)$

$\qquad\qquad \phi_2(x) = \cosh \theta_2(x)\, \sin \theta_3(x)$

$\qquad\qquad \phi_3(x) = \sinh \theta_2(x)\, \sin \theta_3(x)$

$\qquad\qquad \phi_4(x) = \sinh \theta_2(x)\, \cos \theta_3(x)$

Now $h_1(x, t) = 2(R_e H_1(x)) \cos \omega t - 2(i \, m \, H_1(x)) \sin \omega t$

or $\quad h_1(x, t) = 2 \left[H_1(0)\phi_1(x) - 1/\omega b \, Q_1(0) (\theta_2 \, \phi_2(x) + \theta_3 \, \phi_4(x)) \right]$

$\cos \omega t - 2 \left[H_1(0)\phi_3(x) + 1/\omega b \, Q_1(0) (\theta_2 \, \phi_4(x) - \theta_3 \, \phi_2(x)) \right] \sin \omega t$

$$\ldots \text{(3.24)}$$

$$H(x, t) = h_0(x) + h_1(x, t) \qquad \ldots \text{(3.25)}$$

where $h_0(x)$ is the mean level of water at various sections and $h_1(x, t)$ is given by eqn. (3.24). Equation (3.25) is the equation for the level of water with respect to channel bottom.
Similarly the expression for

$$Q_1(x) = \left[C_1(0) \, \phi_1(x) + \frac{\omega b}{G} H_1(0) \ (\theta_2 \, \phi_2(x) - \theta_3 \, \phi_4(x)) \right]$$

$$+ i \left[Q_1(0) \, \phi_3(x) - \frac{\omega b}{G} H_1(0) \ (\theta_2 \, \phi_4(x) + \theta_3 \, \phi_4(x)) \right]$$

where, $\qquad G = \theta_2^2 + \theta_3^2 \qquad \ldots \text{(3.26)}$

$$Q_1(x, t) = 2 \left[Q_1(0) \, \phi_1(x) + \frac{\omega b}{G} H_1(0) \ (\theta_2 \, \phi_2(x) - \theta_3 \, \phi_4(x)) \right]$$

$$\cos \omega t - 2 \left[Q_1(0) \, \phi_3(x) - \frac{\omega b}{G} H_1(0)(\phi_4(x) \theta_2 + \theta_3 \, \phi_2(x)) \right] \sin \omega t$$

$$\ldots \text{(3.27)}$$

$$Q(x, t) = q_0(x) + Q_1(x, t) \qquad \ldots \text{(3.28)}$$

where $q_0(x)$ is the mean discharge at various sections and $Q_1(x, t)$ is given by eqn. (3.27).

LITERATURE CITED

1. Pillsbury, B. (1956). *Tidal Hydraulics*. Corps of Engineers, U.S. Army. Washington, D.C.
2. Perroud, Paul (1958). The propagation of tidal waves into channels of gradually varying cross-section. Institute of Engineering Research, Berkeley, California, October, 1958, series 89, no. 3.
3. Otter, J.R.H. (1960). Tidal Flow Computations. *The Engineer*, January 29, 1960, pp. 177–182.
4. Chattergee, S. (1968). Tidal flow in variable width channel. M.E. Thesis, Bengal Engineering Collage, Sibpur, Howrah.

APPENDIX: FOURIER DEVELOPMENT OF COEFFICIENTS

$$h(x, t) = h_1(x) \cos (\omega t + \alpha_1 (x)) \qquad \text{.... (3A.1)}$$

$$Q(x, t) = q_1 (x) \cos (\omega t + \beta_1 (x)) \qquad \text{.... (3A.2)}$$

where $h_1(x)$, $q_1(x)$ are positive.

Since the channel cross-sections are rectangular, $b(x)$ is independent of h:

$$b(x) = b_0(x), \ A_0(x) = b(x)h_0(x) \qquad \text{.... (3A.3)}$$

where $h(x)$ is the mean depth of water. The cross-sectional area of the stream bed is

$$A = b(h_0 + h)$$

$$A(x, h) = A_0(x) + b(x) h_1(x) \cos \theta \qquad \text{.... (3A.4)}$$

Further development of the coefficients of eqn. (3.7) will be as follows:

$$1/(gA (x, h)) = m_0(x) + m_1(x) h_1(x) \cos \theta \qquad \text{.... (3A.5)}$$

where

$$m_0(x) = 1/(2\pi g) \int_{-\pi}^{\pi} 1/(A (x, h) \, d\theta$$

$$m_1(x) \ h_1(x) = 1/g\pi \int_{-\pi}^{\pi} \cos \theta \ d\theta/(A (x, h))$$

Substituting eqn (3A.4) the expression for

$$m_0(x) = (1/g A_0)\left[(1 - z)^{-1/2}\right] \qquad \text{.... (3A.6)}$$

where $z = (h_1/h_0)^2$ and $m_1(x) \ h_1(x) = 2/(g A_0 z) \left[1 - (1 - z)^{-1/2}\right](h_1/h_0)$

or

$$m_1(x) = 2/(gA_0 z \ h_o) \left[1 - (1 - z)^{-1/2}\right] \qquad \text{.... (3A.7)}$$

By expanding the expressions of $m_0(x)$ and $m_1(x)$ and making simplification, one obtains

$$m_0(x) = 1/gA_0 \left(1 + h_1^2 /2h_0^2\right)$$

and

$$m_1(x) = -1/(gA_0h_0) \left(1 + 3/4 \ h_1^2 /h_0^2\right) \qquad \text{.... (3A.8)}$$

In the term $\left(-\dfrac{b}{gA^2} Q \ dh/dt\right)$, the coefficient (b/gA^2) is usually less important than the coefficients of other terms in the dynamic equation. The variation of A with time need not be considered, so one can write

$$b/(gA^2) = 1/(gA_0^2)$$

Due to the important variation of the term $(C^2RA^2)^{-1}$, the expression of this coefficient is considered to be the third Fourier term. In a wide channel R is approximately equal to h:

$$1/(C^2RA^2) = 1/(C^2A^2h); \text{ let } r = 1/(C^2hA^2)$$

$$r = r_0(x) + r_1h_1(x) \cos\theta_3 + 1/2 \ r_2h_1^2(x) \cos 2\theta$$

where
$$r_0(x) = 1/2\pi C \int_{-\pi}^{\pi} d\theta/(hA^2); r_1(x) h_1(x) = 1/\pi C^2 \int_{-\pi}^{\pi} \cos\theta \frac{d\theta}{(hA^2)}$$

$$1/2r_2(x)h_1^2(x) = 1/\pi C^2 \int_{-\pi}^{\pi} \cos 2\theta \ d\theta/(h A^2) \qquad \text{.... (3A.9)}$$

After substitution of the expression for A, h and integration, the expressions for

$$r_0(x) = \frac{1}{C^2 A_0^2 h_0} \left(1 + 3z + 5 \times 5/8 \ z^2\right)$$

$$r_1(x) = - \frac{3}{C^2 A_0^2 h_0^2} \left(1 + 2 \times 1/2 \ z + 4 \times 3/8 \ z^2\right)$$

$$r_2(x) = \frac{6}{C^2 A_0^2 h_0^3} \left(1 + 2 \times 1/2 \ z + 4 \times 3/8 \ z^2\right) \text{.... (3A.10)}$$

where $z = (h_1/h_0)^2$

Fourier development of factor $Q|Q|$

Herein Fourier development of the quasi-quadrative resistance term $Q|Q|$ is made considering only one-dimensional flow. The Fourier terms for $Q|Q|$ will be:

$$Q|Q| = \sum_{n=1}^{\infty} (c_n \cos n\omega t + d_n \sin n\omega t) \qquad \text{.... (3A.11)}$$

where
$$c_n = \frac{1}{\pi} \int_{-\pi}^{\pi} Q|Q| \cos n\omega t \ d \ (\omega t)$$

$$d_n = \frac{1}{\pi} \int_{-\pi}^{\pi} Q|Q| \sin n\omega t \ d \ (\omega t)$$

Alternately $Q|Q|$ can be expressed as

$$Q|Q| = \sum_{n=1}^{\infty} \left(c_n \, Q_n \, e^{in\omega t} + c_{-n} \, Q_{-n} \, e^{-in\omega t} \right) \qquad \dots \text{(3A.12)}$$

or

$$Q|Q| = c_1 \, Q_1^{i\omega t} + c_{-1} \, Q_{-1} \, e^{-i\omega t}$$

where $c_1 Q_1 = 1/2\pi \int_{-\pi}^{\pi} Q|Q| e^{-i\omega t} \, d(\omega t)$ and $c_{-1} \, Q_{-1}$ is determined by the conjugate integral of $c_1 Q_1$

After evaluation of the integral, $c_1 Q_1$ can be expressed as

$$c_1 Q_1 = 16/3\pi \, Q_1 |Q_1| \qquad \dots \text{(3A.13)}$$

and the complex conjugate $c_{-1} Q_{-1} = 16/3\pi \, Q_{-1} |Q_1|$.

The basic harmonic terms in the development of $Q |Q|$ can be expressed in real terms as:

$$Q|Q| = 16/3\pi \, |Q_1| \left(Q_1 \, e^{i\omega t} + Q_{-1} \, e^{-i\omega t} \right)$$

$$= 8/3\pi \, q_1^2 \cos (\omega t + \beta_1). \qquad \dots \text{(3A.14)}$$

CHAPTER 4

FINITE DIFFERENCE METHOD FOR SOLVING TIDAL FLOW PROBLEMS

4.1 INTRODUCTION

The basic tidal equations are non-linear partial differential equations for which closed form solutions are not available except for very simplified ones. Generally, finite difference methods are adopted for solving practical tidal flow problems. There exist many books such as [2] in which the subject-matter has been dealt with extensively. Here only one solution technic adopted for solving one-dimensional tidal flow problems is discussed.

4.2 MASS AND MOMENTUM EQUATIONS

The St. Venant equations given below were derived on the basis of the following assumptions:

(i)   the water is incompressible and homogeneous, i.e., without significant variations in density;

(ii) the bottom slope is small;

(iii) the wavelengths are large compared to the water depths. This ensures that the flow everywhere can be regarded to have a direction parallel to the bottom, i.e., the vertical acceleration can be neglected and a hydrostatic pressure variation along the vertical can be assumed;

(iv) the flow is subcritical.

The equations are respectively as follows:

$$\frac{\partial Q}{\partial x} + b\,\frac{\partial h}{\partial t} = 0 \qquad\qquad \dots (4.1)$$

$$\frac{\partial Q}{\partial t} + \frac{\partial}{\partial x}\left(\alpha\,\frac{Q^2}{A}\right) + gA\,\frac{\partial h}{\partial x} + \frac{g\,Q|Q|}{C^2 AR} = 0 \quad \dots (4.2)$$

where A is the flow cross-sectional area (m^2); b–total width of the channel (m); C–Chezy's roughness coefficient $(m^{1/2}/s)$; g–acceleration due to gravity (m/s^2); h–stage above horizontal reference level (m); Q–discharge (m^3/s); R–hydraulic radius (m); α–momentum distribution coefficient.

4.3 SOLUTION TECHNIC

4.3.1 Numerical Scheme

The independent variables chosen for the solutions of the basic eqns. (4.1) and (4.2) are respectively Q and h. The method of solution is a fully centred finite difference scheme. The formulation in finite difference form has been done in a computational net which consists of alternating Q and h. In Figure 4.1, Q is defined as a concept in the positive x-direction. The distance between computational points (Δx_j) is free to vary from point to point. The solution method outlined is known as the Abbott and Ionescu [1] scheme.

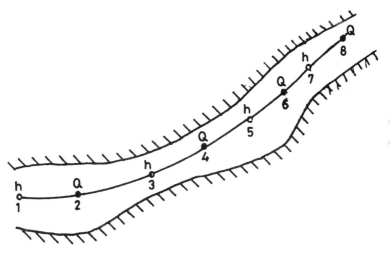

Fig. 4.1   River section with computational net

4.3.2 Continuity Equation

In the continuity equation Q occurs with a derivative with respect to dx. As such it can be centred in an h-point (Fig. 4.2). The derivatives can be expressed on time level $(n + 1/2)$ as follows:

$$\frac{\partial Q}{\partial x} = \frac{1}{\Delta 2x_j}\left[\frac{1}{2}\left(Q_{j+1}^n + Q_{j+1}^{n+1}\right) - \frac{1}{2}\left(Q_{j-1}^n + Q_{j-1}^{n+1}\right)\right] \quad \dots\ (4.3)$$

$$\frac{\partial h}{\partial t} = \frac{1}{\Delta t}\left(h_j^{n+1} - h_j^n\right) \quad\quad \dots\ (4.4)$$

and b is approximated as $b_j^{n+1/2}$ .... (4.5)

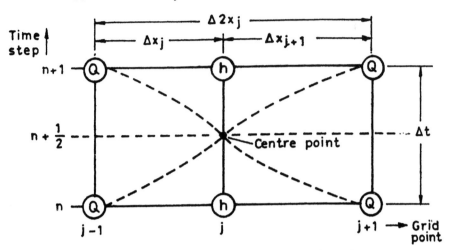

Fig. 4.2 Centering of continuity equation in Abbott scheme

Substituting eqns. (4.3, 4.4 and 4.5) in eqn. (4.1) furnishes formulation of the following form

$$\alpha_j^c \, Q_{j-1}^{n+1} + \beta_j^c \, h_j^{n+1} + \gamma_j^c \, Q_{j+1}^{n+1} = \delta_j^c \qquad \text{.... (4.6)}$$

where

$$\alpha_j^c = - \frac{\Delta t}{2(\Delta 2 x_j)}; \beta_j^c = b_j^{n+1/2}; \; \gamma_j^c = \frac{\Delta t}{2(\Delta 2 x_j)};$$

$$\delta_j^c = h_j^n \, b_j^{n+1/2} - \frac{\Delta t}{2(\Delta 2 x_j)} \left[Q_{j+1}^n - Q_{j-1}^n \right].$$

4.3.3 Momentum Equation

The momentum eqn. (4.2) is centred around Q-points and is shown in Fig. 4.3. The derivatives are expressed in the following manner.

$$\frac{\partial Q}{\partial t} = \frac{1}{\Delta t} \left[Q_{j+1}^{n+1} - Q_{j+1}^n \right]; \; \frac{\partial}{\partial x} \left(\alpha \frac{Q^2}{A} \right)$$

$$= \frac{1}{\Delta 2 x_j} \left[\left(\alpha \frac{Q^2}{A} \right)_{j+1}^{n+1/2} - \left(\alpha \frac{Q^2}{A} \right)_{j-1}^{n+1/2} \right] \qquad \text{.... (4.7)}$$

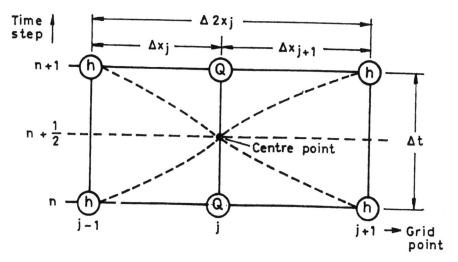

Fig. 4.3 Centering of momentum equation in Abbott scheme

$$gA\,\frac{\partial h}{\partial x} = gA_j^{n+1/2}\,\frac{1}{\Delta 2x_j}\left[\left(h_{j+1}^{n+1} + h_{j+1}^n\right) - \left(h_{j-1}^{n+1} + h_{j-1}^n\right)\right] \quad \dots \ (4.8)$$

$$\frac{gAQ|Q|}{C^2\,A^2\,R} = g\left(\frac{A}{C^2\,A^2\,R}\right)_{j+1}^{n+1/2}\left[fQ_{j+1}^n\,Q_{j+1}^{n+1} + (1-f)Q_{j+1}^n\,Q_{j+1}^n\right] \quad \dots$$

(4.9)

where

$$\Delta 2x_j = x_{j+1} - x_{j-1}, \quad \Delta 2x_{j+1} = x_{j+2} - x_j$$

and f = weighting coefficient.

In this scheme stage and discharge Q are not computed at the same points; the values of Q^2 in the $\dfrac{\partial}{\partial x}\left(\alpha\,\dfrac{Q_2}{A}\right)$ term for the equation have to be evaluated at the points, j and $j + 2$. They are interpolated from their values at neighbouring Q-points with a weighting defined by storage. The parameter f in the resistance term has been introduced because of the possible rapid variation and reversal of. discharge. The solution of the equations above requires iterations where f is taken equal to 1 in the first iteration, while in the later iterations it is given by the value vide eqn. (4.10).

When all the derivatives are substituted, the momentum equation can be written as, vide eqn. (4.11). In the eqn. (4.10) the coefficients with superscript $n + 1/2$ are considered as known functions,

$$f = \frac{\left| Q_j^{n+1/2} \right| Q_j^{n+1/2} - Q_j^n \left| Q_j^n \right|}{Q_j^n \left(Q_j^{n+1} - Q_j^n \right)} \qquad \text{.... (4.10)}$$

of flow variables computed at time level $n\Delta t$:

$$\alpha_{j+1}^m h_j^{n+1} + \beta_{j+1}^m Q_{j+1}^{n+1} + \gamma_{j+1}^m h_{j+2}^{n+1} = \delta_{j+1} \qquad \text{.... (4.11)}$$

where α^c, β^c, γ^c, δ^c α^m, β^m, γ^m, δ^m, are known functions of flow variables. The equations 4.6 and 4.11 are two linear algebraic equations in terms of values Q_{j-1}, h_j, Q_{j+1}, and h_{j+2}. For each pair of points $(j-1, j)$, $(j+1, j+2)$ a system of equa-tions can be established. The coefficients $b_j^{n+1/2}$, $A_{j+1}^{n+1/2}$ and $(A/C^2A^2 R)_{j+1}^{n+1/2}$ are evaluated during iterations. Hence it is to be noted that the coefficients A_{j+1} and $(A/C^2 A^2 R)_{j+1}$ are to be evaluated at Q points where water stages are unknown. So these functions have to be interpolated. As a first approxi-mation consider $b^{n+1/2} = b^n$, $A^{n+1/2} = A^n$, $(C^2A^2R)^{n+1/2} = (C^2A^2R)^n$. The equations when solved furnish first approximations to the variables h_j^{n+1}, Q_{j+1}^{n+1}, $(j = 1, ...n)$ which means second approximations to the coefficients:

$$A^{n+1/2} = \frac{(A^n + A^{n+1})}{2}, (C^2A^2R)^{n+1/2} = \frac{(C^2A^2R)^n + (C^2A^2R)^{n+1}}{2} \quad \text{etc.}$$

Substituting the coefficients in eqns. (4.3) to (4.5) and (4.7) to (4.9) and carrying out another iteration leads to new second approximations to the system for h_j^{n+1}, Q_{j+1}^{n+1}, $j = 1, 2...n$. Satisfactory results are obtained with two iterations at every time step.

4.4   DOUBLE-SWEEP ALGORITHM

It can be seen from eqns. (4.6) and (4.11) that the continuity and momentum equations can be formulated in a similar form. Using the general variable Z, (which becomes h in grid points with odd numbers and Q in grid points with even numbers) the general formulation will be.

$$\alpha_j Z_{j-1}^{n+1} + \beta_j Z_j^{n+1} + \gamma_j Z_{j+1}^{n+1} = \delta_j. \qquad \text{.... (4.12)}$$

Each unknown to the unknown in the neighbouring points can be related by an equation of the form

$$Z_{j-1}^{n+1} = E_j Z_j^{n+1} + F_j \qquad \text{.... (4.13)}$$

where E_j and F_j are quasi-constants,

From the above two equations one obtains the following set of recursive formulae:

$$E_j = \frac{-\gamma_{j-1}}{\beta_{j-1} + \alpha_{j-1} E_{j-1}} \qquad \dots\ (4.14)$$

$$F_j = \frac{\delta_{j-1} - \alpha_{j-1} F_{j-1}}{\beta_{j-1} + \alpha_{j-1} E_{j-1}} \qquad \dots\ (4.15)$$

The complete set of equations for a system of N grid points with R external boundaries will provide N-R linear relations of the general form. At the boundaries other R known values or known relations between the independent variables will form the missing R equations. The relations at the external boundaries are used to initiate the so-called double-sweep algorithm which is based on successive use of eqns. (4.13) and (4.14, 4.15).

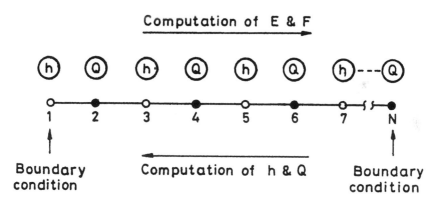

Fig. 4.4 Solution by double-sweep method

Consider now a scheme for computation as shown in Fig. 4.4. For N grid points the two R external boundaries will provide $(N - 2)$ linear equations of the general form. At the boundaries the two known R values or known relations between the independent variables will provide for the missing two equations. The relations at the external boundaries are used to initiate 'the double-sweep' algorithm, which is based on successive use of the equations for E_j and F_j. If the boundary conditions at the channel end where the sweep is initiated is, for instance, a constant boundary with the value h, the sweep is initiated by setting: etc.

$$h_{j-1} = 0 \cdot Q_j + h_j \qquad \dots\ (4.16)$$

40

(h_{j-1} is independent of Q_j) which together with eqn. (4.13) gives

$$E_j = 0, \quad F_j = h_1$$

From these starting values and with successive use of equation for E_j and F_j, E and F can be determined for every grid point. This is called the E-F sweep.

Now $E_2 = 0$, $F_2 = h_1$ for grid point 2, so for grid point 3

$$E_3 = \frac{-\gamma_2}{\beta_2 + \alpha_1 E_2} \qquad \dots \text{(4.17)}$$

$$F_3 = \frac{\delta_2 - \alpha_2 F_2}{\beta_2 + \alpha_2 E_2} \qquad \dots \text{(4.18)}$$

The values of the coefficients to be used in the above expressions are those obtained from momentum equations. Proceeding in this way the values of E and F are evaluated for all the remaining grid points. The boundary condition at the other end, if it is for $Q = 0$, can be used to initiate a backward sweep, where all the unknown h and Q can be evaluated by successive use of eqn. (4.13), i.e., $h - Q$ sweep. In this way one computational cycle is carried out. The cycle is then repeated with more accurate values of the coefficient until the desired accuracy is obtained, usually after two cycles. A software is given in the Appendix for solving tidal propagation problems.

LITERATURE CITED

1. Abbott, M.B and Ionescu, F., (1967), On the numerical computation of nearly horizontal flows. J. *Hydraulic Research*, 5(2): 97-117.
2. Cunge, J.A., Holley. F. M. and Verwey, A. (1980). *Practical Aspects of Computational River Hydraulics*. Pitman Advanced Publishing Company Ltd., London.

APPENDIX: PROGRAM FOR TIDAL PROPAGATION
ANALYSIS IN A NARROW, WELL-MIXED TIDAL REACH

```
      PARAMETER (NRIN = 12, NOUT = 13)
      PARAMETER (IDX = 70)

      COMMON/COMPG/NX, DX (IDX)
      COMMON/BOUND/T1(50), T2(50), TL(50), TQ(50), NCD, NCU

      DIMENSION TIDL (IDX), TIDQ (IDX), TIDV (IDX), TIDA (IDX)
      DIMENSION STADIS (IDX), AMAN (IDX), BW (IDX), HR (IDX)
      DIMENSION ALPHA (IDX), BETA (IDX), GAMMA (IDX), DELTA (IDX)
      DIMENSION P (IDX), R (IDX), S(IDX), T (IDX), TIDLQ (IDX), CHEZ (IDX)
C
C  INPUT FILES AND OUTPUT FILES
C
      OPEN (NRIN, FILE = 'TIDAL. IN', STATUS = 'OLD')
      OPEN (NOUT, FILE = 'TIDAL. OUT', STATUS = 'UNKNOWN')
C
C  SIMULATION INPUT
C
      READ (NRIN,*)
      READ (NRIN,*) TIDPERIOD, DT
      READ (NRIN,*)
      READ (NRIN,*) MAXNCY
      READ (NRIN,*)
      READ (NRIN,*) COURANT
      READ (NRIN,*)
      READ (NRIN,*) EPSL, EPSQ
C
C  GRID PARAMETERS
C
      READ (NRIN,*)
      READ (NRIN,*) NX
      READ (NRIN,*) (DX (I),I = 1, NX)
      ALENGTH = DX (NX)
C
C  OUTPUT RESULTS
C
      READ (NRIN,*)
      READ (NRIN,*)DUMPT
      READ (NRIN,*)
      READ (NRIN,*) NS
      READ (NRIN,*) (STADIS (I),I = 1, NS)
C
C  BED RELATING INPUT (MANNING'S ROUGHNESS COEFFICIENTS).
C
```

```
      READ (NRIN,*)
      READ (NRIN,*) (CHEZ(I), I = 2, NX, 2)
      READ (NRIN,*)
      READ (NRIN,*) DBCOND, UBCOND
C
C     DOWNSTREAM  END  (SEA  MOUTH)
C
      READ (NRIN,*)
      READ (NRIN,*) NCD
      READ (NRIN,*) (T1(I),I = 1,NCD)
      DO I = 1, NCD
        T1(I) = T1(I)*3600.0
      ENDDO
      READ (NRIN,*)(TL(I),I=1,NCD)
C
C     UPSTREAM  END  (LANDWARD)
C
      READ (NRIN, *)
      READ (NRIN, *) NCU
      READ (NRIN,*)(T2(I),I=1,NCU)
      DO I = 1,NCU
        T2(I) = T2(I) * 3600.0
      ENDDO
      READ (NRIN,*) (TQ(I),I=1,NCU)
C
C     CONSTANT  PARAMETERS
C
      READ (NRIN,*)
      READ (NRIN,*)GRA
C
C     CALCULATING HYDRAULIC PARAMETERS AT RIVER CROSS-SECTIONS
C
      CALL CROSS
C
C     INITIAL  CONDITIONS  (ASSUMED)
C
      READ (NRIN,*)
      READ (NRIN,*) (TIDL (I),I = 1, NX,2)
      READ (NRIN,*) (TIDQ (I),I = 2, NX,2)
      CLOSE (NRIN)
C
C     EQUATIONS  OF  SECTIONS
C
      GDT = GRA* DT
C
C     COMPUTATIONAL  CYCLE  STARTS  FROM  HERE
      WRITE (NOUT, *) 'RESULTS'
      NCYCLE = 1
310   WRITE (*,*)' TIDAL CYCLE NO =',NCYCLE
```

```
C
C     COMPUTATIONAL TIME STARTS FROM HERE
C
      TIME–NSEC = 0.0
305   CONTINUE
      WRITE (*,*)'TIME = ', TIME–SEC
C
C     BOUNDARY CONDITIONS
C

C
C     DOWNSTREAM BOUNDARY CONDITIONS
C
      CALL BOUND VAL (TIME SEC, DBC, UBC)
      ITER = 1
100   CONTINUE
        GO TO (21, 22, 23) DBCOND.
21    TIDL(1) = DBC
        TIDLQ (1) = DBC
        GO TO 24
22    TIDQ (1) = DBC
        TIDLQ(I) = DBC
        GO TO 24
23    TIDV (1) = DBC
        TIDLQ (1) = DBC
24    CONTINUE
C
C     UPSTREAM BOUNDARY CONDITIONS
C
      GO TO (31, 32, 33) UBCOND
31    TIDL (NX) = UBC
        TIDLQ (NX) = UBC
        GO TO 34
32    TIDQ (NX) = UBC
        TIDLQ (NX) = UBC
        GO TO 34
33    TIDV (NX) = UBC
        TIDLQ (NX) = UBC
34    CONTINUE
C
C     CALCULATING THE HYDRAULIC PARAMETERS FOR INITIAL CONDI-
      TIONS
C
      DO J = 2, NX-2, 2
      TIDL(J) = 0.5* (TIDL(J-1) + TIDL (j+1))
      ENDDO
      TIDL (NX) = TIDL (NX – 1)

      CALL WEL_AREA (TIDL,TIDA)
      CALL WEL_STWH (TIDL, BW)
```

```
      CALL_WEL HRAD (TIDL,HR)
C
C     CONTINUITY  EQUATION
C
      DO J = 3, NX-1, 2
      DELX2 = DX(J+1)-DX(J-1)
      D1 = BW(J)
      COEFF1= 1./(2*DELX2)
      COEFF2 = BW(J)/DT

      ALPHA (J) = -COFF 1
      BETA   (J) =   COFF 2
      GAMMA       (J) =   COFF1
      DELTA (J) =   COFF2*TIDL(J)-COFF1*(TIDQ(J+1)-TIDQ(J-1).
      ENDDO
C
C     DYNAMIC  EQUATION
C
      ALPHA1 = 1.0
      BETA1 = 1.0
      DOJ = 2, NX-2, 2
      IF (ITER.EQ. 1) THEN
      TIDLQ(J) = TIDQ(J)
      ENDIF
      DELX2  = DX(J+1)-DX(J-1)
      GA     = GRA*TIDA(J)
      GA2    = GRA*TIDA(J)*TIDA(J)
      GA3    = GRA*TIDA  (J)*TIDA(J)*TIDA(J}
      ABS-BB = ALPHA1*BW(J)+BETA1*BW(J)
      C2A2R  = CHEZ(J)*CHEZ(J)*TIDA(J)*TIDA(J)*HR(J)
      TIDQ2  = SQRT((TIDQ(J)*TIDLQ(J))**2)

      D2     = ALPHAI/GA
      D3     = TIDQ(J)/GA2*ABS-BB
      D4     = (1-BETA1*TIDQ(J)*TIDQ(J)*BW(J)/GA3)
      D5     = 1./C2A2R

      SLOPE  = 0.0
      FRICT  = D5/(4*C2A2R)*TIDQ2*TIDQ(J)

      ALPHA(J)  = - D3/(2*DT) -        D4/(2*DELX2)
      BETA (J)  =      D2/DT +        D5/4.*TIDQ2
      GAMMA(J)=  - D3/(2*DT) +        D4(2*DELX2)
      DELTA (J) =     D2/DT* TIDQ(J)-D3/(2*DT)*(TIDL(J+1) + TIDL (J-1))
                    - D4/(2*DELX2)*(TIDL(J+1) - TIDL(J - 1) - SLOPE-FRICT
      ENDDO
C
C     SOLUTION  STRATEGY
C
```

```
      S(I)        = 0.0
      T(I)        = TIDL (1)
      DO J        = 2, NX-2, 2
      PRDENU      = BETA (J) -S (J-1)* ALPLHA(J)
      P(J)        = GAMMA (J)/PRDENU
      R(J)        = (DELTA (J) -T(J-1)*ALPHA(J))/PRDENU
      STDENU      = (BETA(J+1)-ALPHA(J+1) *P(J))
      S(J+1)      = GAMMA(J+1))STDENU
      T(J+1)      = (DELTA(J+1)-R(J)*ALPHA(J+1))/STDENU  ENDDO
      ENDDO

      DO J        = NX-1,2,-2
      TIDLQ(J)    = T(J)-S(J)*TIDLQ(J+1)
      TIDLQ(J-1) = R(J-1)-P(J-1)*TIDLQ(J)
      ENDDO

      IF          (NCYCLE. EQ. MAXNCY) THEN
                  WRITE (NOUT,*) TIME-SEC
                  WRITE (NOUT,15) (TIDLQ(I),I=1,NX-1,2)
                  WRITE (NOUT,15) (TIDLQ(I),I=2,NX,2)
      ENDIF
  15 FORMAT (8(1X,F8.1)
C
C     CONVERGENCY OF FLOW VARIABLES AT THE SAME TIME STEP
C
      IF (ITER .LT. 2) THEN
            DO I = 1, NX-1,2
            TIDL(I) = TIDLQ(I)
            TIDQ (I) = TIDLQ (I+1)
            ENDDO
            ITER = ITER + 1
            GO TO 100
      ENDIF
      IF (TIME-SEC.LT.TIDPERIOD) THEN
      TIME-SEC = TIME -SEC+DT
      GO TO 305
      ENDIF
      IF (NCYCLE . LT. MAXNCY) THEN

      NCYCLE = NCYCLE + 1
      GO TO 310
      ENDIF
C
C     FORMAT STATEMENTS
C
      STOP
      END
```

46

```fortran
SUBROUTINE WEL_HRAD (WSE,VARB)
PARAMETER (IDX = 70)
COMMON/COMPG/NX,DX(IDX)
COMMON/COEFF/A0(100), A8(100),B0(100), B8(100)
DIMENSION WSE (IDX),VARB(IDX)

DO I = 1,NX
COB = (B8(I)-B0(I))*WSE(I)/8.+ B0(I)
COA = (A8(I)-A0(I))*WSE(I)/8. +A)(I)
VARB (I) = COA/COB.
ENDDO
RETURN
END

SUBROUTINE WEL_AREA (WSE,VARB)

PARAMETER (IDX = 70)
COMMON/COMPG/NX,DX(IDX)
COMMON/COEFF/A0(100), A8(100), B0(100), B8(100)
DIMENSION WSE (IDX), VARB(IDX)

DO I = 1,NX
   COA = (A8(I)-A0(I))*WSE(I)/8.+A0(I)
   VARB(I) = COA
ENDDO
RETURN
END

SUBROUTING WEL_STWH(WSE,VARB)

PARAMETER (IDX = 70)
COMMON/COMPG/NX,DX(IDX)
COMMON/COEFF/A0(100), A8(100), B8(100).
DIMENSION WSE (IDX),VARB(IDX)

DO I = 1,NX
   COB = (B8 (I)-B0(I))*WSE(I)/8.+B0(I)
   VARB(I) = COB
ENDDO
RETURN
END

SUBROUTINE BOUND_VAL(TIME_SEC, BOUL, BOUD)

COMMON/BOUND/T1(50), T2(50), TL(50), TQ(50), NCD, NCU

DO K = 1, NCD-1
IF (TIME-SEC.GE.TI(K).AND. TIME_SEC.LE.TI(K+1)) THEN
   DW = T1(K+1)-TI(K)
   DVAR = TL (K+1)-TL(K)
IF (DVAR.GE.O.O) THEN
```

```
      DH1 = TIME_SEC_T1(K)
    WRITE(*,*) DW, DVAR, DH1
    BOUL=DVAR/DW * DHI + TL(K)
    ELSE
    DH2 = TIME_SEC_T1(K+1)
    BOUL = DVAR/DW*DH2 + TL(K+1)
    ENDIF
    GO TO 10
    ENDIF
    ENDDO
 10 CONTINUE

    DO K = 1, NCU-1
    IF (TIME-SEC.GE.T2(K).AND. TIME_SEC.LE.T2(K+1)) THEN
       DW = T2 (K+1) - T2(K)
       DVAR = TQ(K+1)-TQ(K)
       IF (DVAR .GE. 0.0) THEN
           DHI = TIME.SEC - T2(K)
           BOUD = DVAR/DW * DHI + TQ(K)
           ELSE
           DH2 = TIME-SEC-T2(K+1)
           BOUD=DVAR/DW*DH2+TQ(K+1)
           ENDIF
           GO TO 20
    ENDIF
    ENDDO
 20 CONTINUE

    RETURN
    END


    SUBROUTINE CROSS
    PARAMETER (IDX = 70)
    COMMON/COEFF/A0(100), A8(100), B0(100), B8(100)
    COMMON/COMPG/NX,DX(IDX)
       AA10 = 98200.0
       AA18 = 267700.0
       BB10 = 18400.0
       BB18 = 23400.0
       AA20 = 13800.0
       AA28 = 29900.0
       BB20 = 1900.0
       BB28 = 2000.0
    DO I = 1,NX
       DIS = DX(T)
    IF (DX(I) .LE. 85000) THEN
       A0(I) = AA10*EXP(-0.027*DIS/1000.)
       A8(I) = AA18*EXP(-0.027*DIS/1000.)
       B0(I) = BB10*EXP(-0.028*DIS/1000.)
```

48

```
              B8(I)  = BB18*EXP(-0.028*DIS/1000.)
      ELSE
              A0(I) = AA20*EXP(-0.022*DIS/1000.-.85)
              A8(I) = AA28*EXP(-0.016*DIS/1000.-.85)
              B0(I) = BB20*EXP(-0.017*DIS/1000.-.85)
              B8(I) = BB28*EXP(-0.013*DIS/1000.-.85)
      ENDIF
   ENDDO
   RETURN
   END

   TIDPERIOD,DT
   45000.0 100.0
   MAX NO. CYCLE
   4
   COURANT NUMBER
   1.0
   ERROR LIMIT IN WATER LEVELS, DISCHARGE
   0.001 10.0
   NUMBER OF CROSS-SECTION
   62
```

0.0	2440.0	4880.0	7320.0	9760.0	12200.0	14640.0	17080.0
19520.0	21960.0	24400.0	26840.0	29280.0	31720.0	34160.0	36600.0
39040.0	41480.0	43920.0	46360.0	48800.0	51240.0	53680.0	56120.0
58560.0	61000.0	63440.0	65880.0	68320.0	70760.0	73200.0	75640.0
78080.0	80520.0	82960.0	85400.0	87840.0	90280.0	92720.0	95160.0
97600.0	100040.0	102480.0	104920.0	107360.0	109800.0	112240.0	114680.0
117120.0	119560.0	122000.0	124440.0	126880.0	129320.0	131760.0	134200.0
136640.0	139080.0	141520.0	143960.0	146400.0	148840.0		

```
OUTPUT  INTERVAL
  1800.0
NUMBER OF STATIONS
  9
```

0.0	14640.0	34160.0	43920.0	63440.0	92720.0	102480.0	131760.0
146400.0							

CHEZYS ROUGHNESS COEFFICIENTS

100.0	100.0	100.0	90.0	80.0	80.0	75.0	75.0
75.0	75.0	75.0	75.0	75.0	75.0	75.0	75.0
75.0	70.0	60.0	60.0	60.0	60.0	60.0	60.0
60.0	60.0	60.0	55.0	55.0	55.0	55.0	

```
BOUNDARY  CONDITIONS
  1   2

DOWNSTREAM  BOUNDARY  CONDITIONS
  26
```

0.0	0.5	1.0	1.5	2.0	2.5	3.0	3.5
4.0	4.5	5.0	5.5	6.0	6.5	7.0	7.5
8.0	8.5	9.0	9.5	10.0	10.5	11.0	11.5
12.0	12.5	3.40	3.10	2.75	2.45	2.15	1.90
1.60	1.40	1.20	0.95	0.80	0.70	0.60	1.45

2.25	3.20	3.75	4.15	4.50	4.70	4.80	4.70
4.50	4.30	4.10	3.75				

UPSTREAM BOUNDARY CONDITIONS
26

0.0	0.5	1.0	1.5	2.0	2.5	3.0	3.5
4.0	4.5	5.0	5.5	6.0	6.5	7.0	7.5
8.0	8.5	9.0	9.5	10.0	10.5	11.0	11.5
12.0	12.5						
0.0	0.0	0.0	0.0	0.0	0.0	0.0	0.0
0.0	0.0	0.0	0.0	0.0	0.0	0.0	0.0
0.0	0.0	0.0	0.0	0.0	0.0	0.0	0.0
0.0	0.0						

GRAVITY

9.81

INITIAL CONDITIONS

1.43	1.38	1.55	1.76	1.80	1.84	1.74	1.91
1.01	1.23	2.36	2.56	2.62	2.77	2.74	2.94
2.04	2.26	2.25	2.41	3.37	3.33	3.32	3.31
3.32	3.32	3.29	3.32	3.27	3.25	4.28	
0.0	0.0	0.0	0.0	0.0	0.0	0.0	0.0
0.0	0.0	0.0	0.0	0.0	0.0	0.0	0.0
0.0	0.0	0.0	0.0	0.0	0.0	0.0	0.0
0.0	0.0	0.0	0.0	0.0	0.0	0.0	

RESULTS

0.000000E+00

3.4	3.7	3.3	2.9	2.6	2.4	2.4	2.3
2.3	2.2	2.2	2.2	2.2	2.2	2.2	2.2
2.3	2.3	2.4	2.5	2.6	2.7	2.8	2.9
.3.0	3.1	3.1	3.2	3.2	3.3	3.3	
−9753.1	11018.1	9983.2	5797.4	2683.2	1176.4	623.2	429.1
313.5	200.8	94.1	9.1	−46.7	−74.7	−81.0	−72.8
−56.8	−38.1	−35.0	−31.5	−27.8	−24.2	−20.7	−17.4
−14.3	−11.5	−8.8	−6.3	−4.1	−2.0	0.0	

1800.00

3.1	3.5	3.3	2.9	2.6	2.5	2.4	2.3
2.3	2.2	2.2	2.2	2.2	2.2	2.2	2.2
2.3	2.3	2.4	2.5	2.6	2.7	2.8	2.9
3.0	3.1	3.1	3.2	3.2	3.3	3.3	
−12551.8	5593.3	8537.6	5839.0	3015.7	1406.6	719.1	452.0
312.9	199.1	96.5	14.5	−40.5	−69.2	−76.8	−70.0
−55.3	−37.8	−34.8	−31.4	−27.8	−24.2	−20.7	−17.4
−14.3	−11.5	−8.8	−6.3	−4.1	−2.0	0.0	

3600.00

2.8	3.3	3.2	3.0	2.7	2.5	2.4	2.3
2.3	2.2	2.2	2.2	2.2	2.2	2.2	2.2

2.3	2.3	2.4	2.5	2.6	2.7	2.8	2.9
3.0	3.1	3.1	3.2	3.2	3.3	3.3	
−16892.6	1510.8	6730.1	5568.5	3227.2	1615.6	825.0	486.3
318.4	199.2	98.9	19.4	−34.8	−63.9	−72.6	−67.2
−53.8	−37.5	−34.6	−31.3	−27.7	−24.1	−20.6	−17.4
−14.3	−11.4	−8.8	−6.3	−4.1	−2.0	0.0	−56.8
5400.00							
2.5	3.1	3.2	3.0	2.7	2.5	2.4	2.3
2.3	2.2	2.2	2.2	2.2	2.2	2.2	2.2
2.3	2.3	2.4	2.5	2.6	2.7	2.8	2.9
3.0	3.1	3.1	3.2	3.2	3.2	3.3	
−19555.1	−1789.6	4867.9	5058.1	3305.7	1787.5	932.9	529.6
329.8	201.6	101.6	23.9	−29.3	−58.8	−68.6	−64.5
−52.3	−37.1	−34.4	−31.1	−27.6	−24.1	−20.6	−17.4
−14.3	−11.4	−8.8	−6.3	−4.1	−2.0	0.0	
7200.00							
2.2	2.9	3.1	3.0	2.7	2.6	2.4	2.3
2.3	2.3	2.2	2.2	2.2	2.2	2.2	2.2
2.3	2.3	2.4	2.5	2.6	2.7	2.8	2.9
3.0	3.1	3.1	3.2	3.2	3.2	3.3	
−22096.2	−4521.2	3083.5	4391.2	3258.3	1910.8	1034.6	578.8
346.6	206.6	104.9	28.3	−24.2	−54.0	−64.7	−61.8
−50.9	−36.8	−34.2	−31.0	−27.5	−24.0	−20.6	−17.3
−14.3	−11.4	−8.8	−6.3	−4.1	−2.0	0.0	
9000.00							
1.9	2.7	3.0	2.9	2.8	2.6	2.4	2.4
2.3	2.3	2.2	2.2	2.2	2.2	2.2	2.2
2.3	2.3	2.4	2.5	2.6	2.7	2.8	2.9
3.0	3.1	3.1	3.2	3.2	3.2	3.3	
−23312.3	−6739.1	1425.7	3634.5	3103.3	1980.2	1123.1	630.6
368.0	214.3	109.2	32.8	−19.4	−49.4	−60.9	−59.2
−49.5	−36.5	−33.9	−30.8	−27.4	−23.9	−20.6	−17.3
−14.3	−11.4	−8.8	−6.3	−4.1	−2.0	0.0	
10800.0							
1.6	2.6	2.9	2.9	2.8	2.6	2.5	2.4
2.3	2.3	2.2	2.2	2.2	2.2	2.2	2.2
2.3	2.3	2.4	2.5	2.6	2.7	2.8	2.9
3.0	3.1	3.1	3.2	3.2	3.2	3.3	
−25680.2	−8699.2	−87.9	2837.6	2862.8	1995.6	1193.3	681.3
392.7	224.5	114.4	37.3	−14.6	−44.9	−57.3	−56.6
−48.0	−36.1	−33.7	−30.7	−27.3	−23.9	−20.5	−17.3
−14.3	−11.4	−8.8	−6.3	−4.1	−2.0	0.0	
12600.0							
1.4	2.4	2.8	2.9	2.8	2.6	2.5	2.4
2.3	2.3	2.2	2.2	2.2	2.2	2.2	2.2

2.3	2.3	2.4	2.5	2.6	2.7	2.8	2.9
3.0	3.0	3.1	3.2	3.2	3.2	3.2	
−25654.2	−10267.6	−1471.3	2030.4	2557.8	1960.5	1241.6	727.9
419.2	236.9	120.9	42.2	−10.0	−40.6	−53.8	−54.1
−46.7	−35.8	−33.5	−30.5	−27.2	−23.8	−20.5	−17.3
−14.3	−11.4	−8.8	−6.3	−4.1	−2.0	0.0	

14400.0

1.2	2.2	2.7	2.8	2.7	2.6	2.5	2.4
2.3	2.3	2.2	2.2	2.2	2.2	2.2	2.2
2.3	2.3	2.4	2.5	2.6	2.7	2.8	2.9
3.0	3.0	3.1	3.2	3.2	3.2	3.2	
−25920.9	−11452.9	−2690.5	1237.7	2206.9	1880.5	1266.5	767.7
446.1	250.9	128.4	47.5	−5.4	−36.5	−50.3	−51.7
−45.3	−35.5	−33.2	−30.3	−27.1	−23.7	−20.4	−17.2
−14.2	−11.4	−8.8	−6.3	−4.1	−2.0	0.0	

16200.0

0.9	2.1	2.6	2.7	2.7	2.6	2.5	2.4
2.3	2.3	2.2	2.2	2.2	2.2	2.2	2.2
2.3	2.3	2.4	2.5	2.6	2.7	2.8	2.9
3.0	3.0	3.1	3.2	3.2	3.2	3.2	
−27290.3	−12551.8	−3767.9	478.1	1826.7	1762.3	1267.9	798.8
471.8	266.2	136.9	53.2	−0.7	−32.4	−47.0	−49.3
−44.0	−35.1	−33.0	−30.2	−27.0	−23.7	−20.4	−17.2
−14.2	−11.4	−8.8	−6.3	−4.1	−2.0	0.0	

18000.0

0.8	1.9	2.5	2.7	2.7	2.6	2.5	2.4
2.3	2.3	2.3	2.2	2.2	2.2	2.2	2.2
2.3	2.3	2.4	2.5	2.6	2.7	2.8	2.9
3.0	3.0	3.0	3.1	3.2	3.2	3.2	
−26584.6	−13401.9	−4734.0	−242.7	1430.2	1612.8	1246.7	819.7
495.2	281.8	146.3	59.3	4.1	−28.4	−43.8	−47.0
−42.7	−34.8	−32.7	−30.0	−26.9	−23.6	−20.4	−17.2
−14.2	−11.4	−8.8	−6.3	−4.1	−2.0	0.0	

19800.0

0.7	1.8	2.4	2.6	2.7	2.6	2.5	2.4
2.3	2.3	2.3	2.2	2.2	2.2	2.2	2.2
2.3	2.3	2.4	2.5	2.6	2.7	2.8	2.9
3.0	3.0	3.1	3.1	3.2	3.2	3.2	
−25281.8	−13870.7	−5557.5	−917.8	1027.2	1438.6	1204.6	829.8
515.1	297.4	156.2	65.9	9.0	−24.4	−40.6	−44.8
−41.4	−34.4	−32.5	−29.8	−26.8	−23.5	−20.3	−17.2
−14.2	−11.4	−8.8	−6.3	−4.1	−2.0	0.0	

21600.0

0.6	1.7	2.3	2.6	2.6	2.6	2.5	2.4
2.4	2.3	2.3	2.2	2.2	2.2	2.2	2.2
2.3	2.3	2.4	2.5	2.6	2.7	2.8	2.9
3.0	3.0	3.1	3.1	3.2	3.2	3.2	
−24383.7	−14101.3	−6228.0	−1536.5	626.9	1245.8	1143.8	828.8
530.7	312.1	166.5	72.9	14.2	−20.4	−37.5	−42.6
−40.1	−34.1	−32.2	−29.6	−26.6	−23.5	−20.3	−17.1
−14.2	−11.4	−8.8	−6.3	−4.1	−2.0	0.0	

23400.0

1.5	1.7	2.2	2.5	2.6	2.6	2.5	2.4
2.4	2.3	2.3	2.2	2.2	2.2	2.2	2.2
2.3	2.3	2.4	2.5	2.6	2.7	2.8	2.9
3.0	3.0	3.1	3.1	3.2	3.2	3.2	
−5145.3	−12103.1	−6593.5	−2081.7	237.7	1040.2	1066.8	817.0
541.3	325.4	176.8	80.1	19.4	−16.3	−34.4	−40.4
−38.9	−33.7	−32.0	−29.5	−26.5	−23.4	−20.2	−17.1
−14.2	−11.4	−8.8	−6.3	−4.1	−2.0	0.0	

25200.0

2.3	1.8	2.1	2.4	2.6	2.6	2.5	2.4
2.4	2.3	2.3	2.2	2.2	2.2	2.2	2.2
2.2	2.3	2.4	2.5	2.6	2.7	2.8	2.9
2.9	3.0	3.1	3.1	3.2	3.2	3.2	
11561.3	−7591.4	−6222.7	−2469.2	−124.0	828.2	976.4	794.7
546.6	336.8	186.8	87.5	24.8	−12.2	−31.4	−38.3
−37.7	−33.4	−31.7	−29.3	−26.4	−23.3	−20.2	−38.3
−14.1	−11.4	−8.8	−6.3	−4.1	−2.0	0.0	−17.1

27000.0

3.2	2.1	2.2	2.4	2.5	2.6	2.5	2.4
2.4	2.3	2.3	2.3	2.2	2.2	2.2	2.2
2.2	2.3	2.4	2.5	2.6	2.7	2.8	2.9
2.9	3.0	3.1	3.1	3.2	3.2	3.2	
31587.0	−1067.2	−4978.4	−2594.3	−426.3	618.8	875.7	763.0
546.3	346.0	196.3	95.0	30.3	−8.1	−28.3	−36.2
−36.5	−33.1	−31.5	−29.1	−26.3	−23.2	−20.1	−17.1
−14.1	−11.4	−8.8	−6.3	−4.1	−2.0	0.0	

28800.0

3.8	2.5	2.2	2.4	2.5	2.5	2.5	2.4
2.4	2.3	2.3	2.3	2.2	2.2	2.2	2.2
2.3	2.3	2.4	2.5	2.6	2.7	2.8	2.9
2.9	3.0	3.1	3.1	3.2	3.2	3.2	
39362.6	5814.4	−2884.7	−2376.1	−630.3	425.8	769.3	722.9
540.5	352.6	205.0	102.3	35.9	−3.9	−25.3	−34.2

−35.3	−32.7	−31.2	−28.9	−26.1	−23.1	−20.0	−17.0
−14.1	−11.3	−8.7	−6.3	−4.1	−2.0	0.0	

30600.0

4.2	2.8	2.4	2.4	2.5	2.5	2.5	2.4
2.4	2.3	2.3	2.3	2.2	2.2	2.2	2.2
2.3	2.3	2.4	2.5	2.6	2.7	2.8	2.9
2.9	3.0	3.1	3.1	3.2	3.2	3.2	
42942.2	11584.5	−323.7	−1806.6	−701.5	265.7	663.3	676.5
529.4	356.4	212.6	109.4	41.5	0.3	−22.2	−32.1
−34.2	−32.4	−30.9	−28.7	−26.0	−23.0	−20.0	−17.0
−14.1	−11.3	−8.7	−6.3	−4.1	−2.0	0.0	

32400.0

4.5	3.2	2.5	2.4	2.5	2.5	2.5	2.4
2.4	2.3	2.3	2.3	2.2	2.2	2.2	2.2
2.3	2.3	2.4	2.5	2.6	2.7	2.8	2.9
2.9	3.0	3.1	3.1	3.2	3.2	3.2	
45647.0	16264.5	2372.3	−948.0	−623.8	154.2	564.8	626.2
513.7	357.4	218.9	116.1	47.1	4.5	−19.2	−30.0
−33.0	−32.0	−30.7	−28.5	−25.9	−22.9	−19.9	−16.9
−14.1	−11.3	−8.7	−6.3	−4.1	−2.0	0.0	

34200.0

4.7	3.4	2.7	2.5	2.5	2.5	2.5	2.4
2.4	2.3	2.3	2.3	2.3	2.2	2.2	2.2
2.3	2.3	2.4	2.5	2.6	2.7	2.8	2.8
2.9	3.0	3.1	3.1	3.2	3.2	3.2	
44161.2	19621.9	4984.4	116.5	−400.0	102.9	481.5	575.1
494.3	355.6	223.8	122.2	52.5	8.7	−16.1	−28.0
−31.9	−31.7	−30.4	−28.3	−25.7	−22.8	−19.9	−16.9
−14.0	−11.3	−8.7	−6.3	−4.1	−2.0	0.0	

36000.0

4.8	3.7	2.9	2.5	2.5	2.5	2.5	2.4
2.4	2.3	2.3	2.3	2.3	2.2	2.2	2.2
2.3	2.3	2.4	2.5	2.6	2.7	2.8	2.8
2.9	3.0	3.1	3.1	3.2	3.2	3.2	
40623.1	21483.7	7284.5	1292.2	−47.1	118.0	420.1	526.9
472.4	351.2	227.3	127.8	57.7	12.8	−13.1	−26.0
−30.7	−31.3	−30.1	−28.1	−25.6	−22.7	−19.8	−16.9
−14.0	−11.3	−8.7	−6.3	−4.0	−2.0	0.0	

37800.0

4.7	3.8	3.0	2.6	2.5	2.5	2.5	2.4
2.4	2.3	2.3	2.3	2.3	2.2	2.2	2.2
2.3	2.3	2.4	2.5	2.6	2.7	2.8	2.8
2.9	3.0	3.1	3.1	3.1	3.2	3.2	

32043.7	21475.7	9067.8	2476.2	406.4	199.3	385.7	485.0
449.6	344.8	229.4	132.7	62.6	16.9	−10.0	−23.9
−29.6	−31.0	−29.9	−27.9	−25.4	−22.7	−19.7	−16.8
−14.0	−11.3	−8.7	−6.3	−4.0	−2.0	0.0	

39600.0

4.5	3.9	3.2	2.7	2.5	2.5	2.5	2.4
2.4	2.4	2.3	2.3	2.3	2.2	2.2	2.2
2.3	2.3	2.4	2.5	2.6	2.7	2.7	2.8
2.9	3.0	3.1	3.1	3.1	3.2	3.2	

41400.0

4.3	3.9	3.3	2.8	2.6	2.5	2.5	2.4
2.4	2.4	2.3	2.3	2.3	2.3	2.2	2.2
2.3	2.3	2.4	2.5	2.6	2.7	2.7	2.8
2.9	3.0	3.0	3.1	3.1	3.2	3.2	
15411.6	17098.7	10456.7	4424.9	1415.7	531.6	407.5	431.9
407.8	328.2	229.6	140.1	71.5	24.7	−4.1	−19.9
−27.3	−30.3	−29.3	−27.5	−25.2	−22.4	−19.6	−16.7
−13.9	−11.2	−8.7	−6.3	−4.0	−1.9	0.0	

43200.0

4.1	3.8	3.3	2.9	2.6	2.5	2.5	2.4
2.4	2.4	2.3	2.3	2.3	2.3	2.2	2.2
2.3	2.3	2.4	2.5	2.6	2.6	2.7	2.8
2.9	3.0	3.0	3.1	3.1	3.2	3.2	
9618.9	14260.0	10209.5	5037.3	1915.1	754.54	62.1	424.8
391.9	319.4	228.2	142.8	75.4	28.4	−1.2	−17.9
−26.2	−29.9	−29.1	−27.3	−25.0	−22.3	−19.5	−16.7
−13.9	−11.2	−8.7	−6.3	−4.0	−1.9	0.0	

45000.0

3.8	3.7	3.4	3.0	2.7	2.5	2.5	2.4
2.4	2.4	2.3	2.3	2.3	2.3	2.3	2.2
2.3	2.3	2.4	2.5	2.6	2.6	2.7	2.8
2.9	3.0	3.0	3.1	3.1	3.2	3.2	
538.1	10853.3	9511.3	5386.4	2384.0	991.5	540.5	431.7
381.1	311.4	226.2	144.7	78.9	31.9	1.6	−16.0
−25.1	−29.6	−28.8	−27.1	−24.9	−22.2	−19.5	−16.6
−13.9	−11.2	−8.7	−6.3	−4.0	−1.9	0.0	

CHAPTER 5

SEDIMENT TRANSPORT IN TIDAL ENVIRONMENT

5.1 INTRODUCTION

Engineering problems associated with tidal conditions very often require an assessment of the sediment source. Possible sources of sediment within estuary systems can be mentioned as follows:

   i) erosion of land by rivers and streams;
  ii) discharge of waste materials and solid wastes;
 iii) bank erosion;
  iv) long-shore transport or littoral drift;
   v) wind erosion of coastal dunes and drying due to exposure of intertidal shoal;
  vi) erosion of near-shore continental shelf; and
 vii) return of dredge spoil and waste generated due to marine plants and animals.

The most important of the above sources are those due to (i), (iii), (iv) and (v). If dredged materials are dumped close to the estuary entrance or persistent shoaling zones they may also contribute significantly.

Sediment deposits in estuaries consist of various proportions of sand, silt, clay and organic matters. Sand is often found at the seaward end where wave action and currents remove the finer fraction. On the other hand, fine sand, silt, clay and organic matter are found in the upper reaches near the limit of tidal average salinity intrusion.

5.2 SEDIMENT TRANSPORT PROCESSES

The erosion, transport and deposition of sediments in tidal flows depend on the physicochemical properties of bed sediments and the surrounding fluid. Generally sand and silt particles move as solid material but clay particles adhere like flocs which change in size and shape during motion. Studies of sediment transport problems should therefore include details of the size, distribution, density and shape characteristic of individual sediment grains as well as the bulk properties of the total deposit.

5.2.1 Transport due to Tidal Velocities

The transport mechanism during a tidal cycle can be illustrated by the experimental observation of Dillo [1] at the Franzuis Institute, Germany. In an experimental flume 55 m long and 0.6 m in depth sand was used with a flow velocity ranging from 0.5 to 2.0 m/s during different phases of the tidal cycle. The following phenomenon was noticed. Up to an average velocity of 37 cm/s, the entire bed materials remained stationary. With increasing velocity individual particles began to roll. As soon as the flow velocity reached near about 50 cm/s, a whirling rise of the materials was observed with a large quantity of sand coming into suspension. The earlier clear water, with further rise of the velocity over the entire depth of the channel, became muddy with suspended sand. Up to a flow velocity of 0.5 m/s, considerable material remained in suspension. With further reduction in velocity, the materials which had gone into suspension rapidly settled down and the water became clear.

The prototype measurements also confirmed model observations. The bed and suspended materials were collected simultaneously by Kestner [2] from the Lune River during different phases of a tidal cycle. Particle size analysis of the bed and suspended materials showed a continuous exchange of materials between bed and suspensions.

The basic transport process in tidal flow is shown in Figure 5.1. Sediment particles are suspended when the current velocity exceeds a certain critical value. In accelerating flow there is always a net vertical upward transport of sediment particles due to turbulence related diffusive processes which continues as long as the sediment transport capacity exceeds the actual transport rate. The time lag period T is the time period between the moment of maximum flow and the moment of equal transport capacity and actual rate. After this latter moment there is a net downward sediment transport because settling dominates, yielding smaller concentration and transport rates. In case of very fine sediments (silts) or a large depth the settling process can continue during the slack period between a zero transport capacity and the start of a new erosion cycle. Figure 5.1 shows that the suspended sediment transport during decelerating flow is always larger than during accelerating flow. It is possible to derive transport relationship as a function of current velocity. This can be done by linear regression analysis on the logarithmic values of the measured transport. However, not much reliability can be placed on the relations as the transport can be anywhere between 0.5 to 2 times the calculated values. The sediment transport

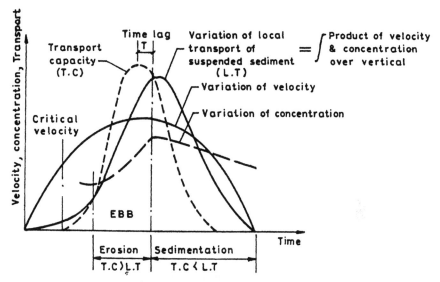

Fig. 5.1 Suspended sediment transport in a tidal flow

formula normally used is that of Kalinske-Kirkham [8]. It can be expressed as follows:

$$S_s = \rho_s \int_a^h (u_z\, c_z)\, dz \qquad \qquad \dots\ (5.1)$$

$$c_z = c_a \exp\left(\frac{-15wz}{u_*h}\right) \qquad \qquad \dots\ (5.2)$$

$$w = \frac{\gamma\omega}{u_*} \qquad \qquad \dots\ (5.3)$$

$$c_a = \frac{A}{\sqrt{\pi}}\left(\frac{1}{2w}\exp(-w^2) \pm \frac{1}{2}\sqrt{\pi} + \int_0^w \exp(-\varepsilon^2)\, d\varepsilon\right) \qquad \dots\ (5.4)$$

$$\varepsilon = \frac{1}{15}\, hu_* = 0.16\, hu_* \qquad \qquad \dots\ (5.5)$$

where A–constant– .0039; a–bed boundary level above the bed; c_a–concentration at the bed; c_z–concentration at a level z above the bed; h–flow depth S_s–suspended load transport per unit width in $kg/m/s$; u_*–shear velocity; γ–constant–2.5; ω–particle fall velocity.

The magnitude of bed load can be determined from the rate of migration of ripples in terms of depth averaged flow velocity u. Normally a power relationship is obtained if the form

$$S_b = ku^n \qquad \qquad \text{.... (5.6)}$$

5.3 SUSPENDED LOAD MEASUREMENTS

In the case of tidal flow, flow is reversible. Estuaries and channels contain a fairly large proportion of fine sand varying from 0.06 to 0.2 mm, which can be readily set in motion by tidal currents.

$$\int_{T_1}^{T_2} g'_m dt - \int_{T_1}^{T_2} g'_c \, dt + \gamma_s \left(A_{D_{T_2}} - A_{D_{T_1}} \right) = 0 \qquad \text{.... (5.7)}$$

Transport mainly occurs in suspension. Therefore measurements of suspended sediments are of great importance. The main objective of study of sediment transport in an estuary is to estimate siltation at public utility structures and within dredged channels through balance of transport in flood and ebb. This may be visualised from the sediment continuity equation adopted to a section of hydraulic measurements indicated by Biswas and Mitra [3] vide eqn 5.7, where g'_m is sediment load in weight per second per unit width entering a section obtained from sampling; g'_c—predicted total bed material load in weight per unit width per second. This is computed from the transport equation; A_D—volume of material between bed and datum over a discrete reach; T_1, T_2—time limit of computation; γ_s—the specific weight of sediment.

$$g'_m = \rho \bar{u} \bar{c} \qquad \qquad \text{.... (5.8)}$$

where ρ–density of fluid; $\bar{u}$–average velocity at time t; $\bar{c}$–average concentration at time t.

In India, Calcutta Port Trust has a system for collection of data in the Hooghly estuary. Let it suffice to mention here that the Hooghly estuary is well-mixed. Vertical salinity gradient is practically absent. A typical time variation of depth averaged velocity, sediment concentration and water level is given in Fig. 5.2.

5.3.1 Computation of Suspended Sediment Load

Generally at a section for a particular tide, observations are made simultaneously at three verticals and at six different depths at each vertical at intervals

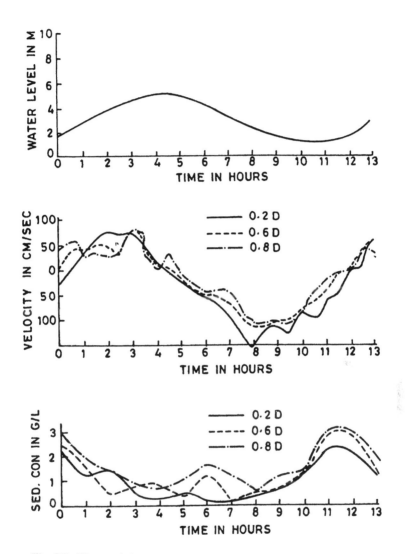

Fig. 5.2 Time variation of water level, velocity and sediment concentration

of one hour throughout the tidal cycle. From the data thus collected it is possible to estimate the suspended sediment discharge across a unit length of the section at the observation vertical or through a unit area at a particular depth in the vertical during a full tidal cycle and to determine the ratio of the ebb discharge to the total sediment discharge.

The directions of velocities for flood or ebb are not always perpendicular to the cross-sectional line, so it is necessary to take the component of

60

velocity along that line. Figure 5.3 shows that the component of ebb velocities should be along 93° and that of the flood velocities along the 273° line. To determine the sediment discharge per unit width at an observation vertical at a particular point of time the velocity components at a right angle to the section at different depths are multiplied by the corresponding concentration. These values are integrated along the depth to find the sediment discharge per unit width for that particular instant. The values thus obtained are integrated along time over an entire tidal cycle. The flood and ebb period sediment discharge are separated. The ratio of ebb volume to the combined flood and ebb volume of sediment discharge yields the ebb predominance. Similarly, by integrating the sediment discharge obtained at a particular depth at different instants of time one can obtain the ebb predominance of sediment discharge at a particular depth. Such routine computations can be done efficiently on a digital computer. A program with dataset and results is enclosed in Appendix I.

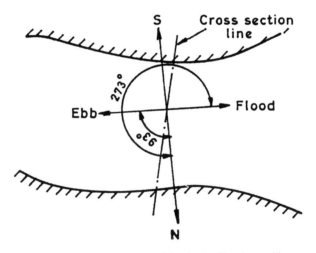

Fig. 5.3 Cross-section showing flood and ebb velocity directions with respect to North

5.4 SAMPLING TECHNICS

Sediment sampling is invariably conducted along with discharge measurements across a cross-section. A pragmatic approach is to select equally spaced segments to permit completion of sampling in one observation. For a tidal river sampling is required to be extended over the complete tidal period, consisting of flood and ebb, preferably every half an hour. Sampling may be point samples or depth integrated. Point samples may be instantaneous or time integrated. Integrated samples may denote summation over depth or time or both. Generally point sampling at 0.6 depth of flow below the surface is adopted. This may be either instantaneous or integrated, as determined by the type of sampler. A depth integrated sampler theoretically samples continuously over the depth of flow by lowering or raising at a uniform rate. It is important to note that in all these samplings, velocities are point integrated. Therefore these are compatible only with point integrated samples. This method decidedly gives approximate results for the overall mean concentration. Multipoint samples are needed to investigate the vertical distribution of suspended sediment concentration. For small values of (ω/ku_*) vertical concentration is negligible, where ω is fall velocity of sediments of different diameters, k–Karman's constant, $u_* = $ shear velocity.

The number of point samples may be determined by the degree of comprehensive information required to be obtained; apparently no standard can be logically laid down for this.

Sampling in tidal rivers is subject to complexities with gradually varied flow and reversal of currents. It is difficult to get a representative depth for sampling as the changing hydrodynamic conditions are so diverse.

However, in alluvial estuaries such as the Hooghly in West Bengal point samples at 0.6 depth below the surface may be adopted. Calcutta Port Trust obtained samples at 0.2, 0.6, 0.8 times depth for differnet stations.

Calcutta Port Trust conducts sediment and discharge observations at suitable site of the Hooghly River. Discharge observation are undertaken on 12 to 14 verticals depending on flood stages. However, sediment samples are collected at only three verticals at 0.6 depth. Satisfactory results have been obtained keeping in view the objective, namely to evaluate the sediment load moving into the tidal estuary from the predominantly non-tidal zone. Vanoni [4] has provided comprehensive field data on vertical distribution of suspended sediments and also average concentration for U.S. rivers.

Due to the water surface continuously rising and falling during the tidal period the specified depth ratios of 0.2, 0.6, 0.8 and the like carry little logic for any point sampling. Perhaps depth-integrated samples may be preferred for such conditions. Calcutta Port Trust sampling results presented in Table 5.1 appear to be satisfactory due to small vertical gradients in concentration within an environment of fine alluvial sediments.

Depth-integrated samples and point-integrated samples may be collected at stream verticals representing areas of equal water discharge in the cross-section or at equally spaced streams vertically in the section. The concentration for the stream cross-section is the mean of concentrations of stream verticals sampled. Again, sample collection would fall more within the domain of judgement than suffer restriction within framed standards. The subtle difference between depth-integrated and point-integrated samples may be appreciated from the following treatment.

In the depth-integrated sampling

$$S_s = \int S_{s_{yz}} \, dA \qquad \qquad \dots \text{(5.9)}$$

$$S_{s_{yz}} = \frac{1}{T} \int_A^D \int f(y, z) \, u(yzt) \, dy \, dt \qquad \dots \text{(5.10)}$$

where, $f(y, z)$ is depth-integrated concentration as a function of depth and position of vertical. This is time averaged; $u(yzt)$–point velocity; S_s–suspended sediment load discharge per unit area; S_s–total suspended load discharge over cross-sectional area A; S_s may be represented by $\overline{C_m}\underline{Q}$, where $\overline{C_m}$ is depth-integrated time average concentration; Q is discharge. $\overline{C_m}, Q$ are measured quantities.

In point-integrated sampling

$$S_{s_{yz}} = \frac{1}{T} \int_0^t C(yzt) \, u(yzt) \, dt \qquad \dots \text{(5.11)}$$

$$S_s = \int_0^A S_s(yz) \, dA \qquad \qquad \dots \text{(5.12)}$$

where $S_s(yz)$ is local sediment flux per unit area;

$C(yzt)$–point concentration at time, t;

$u(yzt)$–velocity at point, t;

T–time period of sampling. The above equation (viz. 5.11) gives eqn. (5.12).

Again $\qquad S_s(yz) = \overline{C}\overline{u} + \overline{C'u'} \qquad \qquad \dots \text{(5.13)}$

where $\overline{C}$ average concentration; $\overline{u}$–everage velocity; $C'u'$–turbulent fluctuations in C, u.

Therefore, measuring C and u does not give the true time sediment flux. Instantaneous samplers or flow-through samplers appear to be preferable, giving an immediate sampling of $S_s(y \, z)$. All these considerations show the niceties involved in all sampling technics. Experience has been varied. For the field it is advisable to proceed on average values permitting the adoption of simple instruments. From conception of the sediment transport in specific cases an optimum sampling technic may emerge. It has to be recognised that in field measurements, sampling and velocity measurements generally remain time averaged. Therefore, guided by the veritable imponderables involved in such measurements it is pragmatic to select a few representative verticals over a cross-section and adopt representative single-point or two-point or three-point/multiple observations to enable determination of an average concentration. Weightage may be given to the concentration in proportion to the current velocities.

5.4.1 Samplers

Ideal technical requirements of a sampler appear in IS and other literature on sedimentation engineering. The prime requirements involve the elements of equality of velocity of sampling and velocity of flow and least disturbance to flow by the sampler immersed. It is indeed difficult to fulfil these ideally. With high velocities of flow when sediment concentration is high, a sampler can rarely be kept in the desired position either vertical or horizontal. Movement of bed forms introduces additional morphological complexities in high floods or rapid flows.

Extensive renovations have gone into the design of samplers. They may be broadly classified as:

i) Instantaneous samplers of the flow through type in the horizontal position. They may be vertical as well.
ii) Bottle samplers or point integrating samplers collecting a sample over a period of time.
iii) Depth integrating samplers collecting a sample over a period of time.
iv) Pump samplers drawing through a pipe.

It may be noted that in all these instruments the criteria enumerated above are satisfied to the extent possible. Although Vanoni [4] categorised 'Instantaneous Samplers' as not suitable for field use, the reasons do not appear to be convincing. The limitations of a bottle sampler are obvious 'Depth' and 'Point' integrators appears to be ideal for use within an environment of high velocities. Pump samplers used by the Calcutta Port Trust in tidal conditions are found to be suitable and appear to give consistent results. Pump samplers may be designed to give a continuous record of sediment concentration at a fixed point or may also provide point-integrated samples. Instantaneous horizontal samplers (Fig. 5.4) used by CPT have given dependable results.

Sediment sampling is always undertaken at the cross-section of discharge observations; otherwise it remains non-informative. A typical sediment sampling recording sheet and processing format are shown in Table 5.1. These formats are in use by CPT. The entire discharge observations may be made on line on a portable data logger using a cassette for further processing. The sediment sampling results may be entered appropriately subsequently. Data retrieval and processing will be guided by use.

64

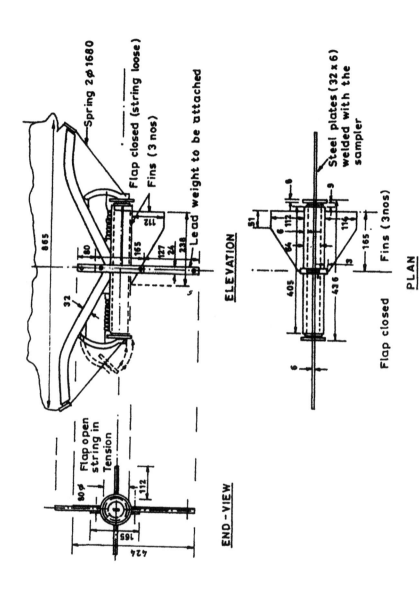

Fig. 5.4 Suspended sediment sampler (dimensions given in mm are approximate)

Table 5.1 Statement of suspended sediment analysis

Date	Discharge in m³/sec.				Sedt.Conc. at 0.6D in gms/lit.			Av. sedt. at 0.6D in gms/lit.	Sedt. flow in tons/day × 10³			Av. total sedt. flow in tons/day × 10³	Amount of water flow in m³/sec. × 10³	Suspended sedi. percent			Suspended sedi. flow in tons/day × 10³			Remarks
	Ll Channel	Mid Channel	Rt Channel	Total discharge in m³/sec.	Ll Channel	Mid Channel	Rt Channel		Ll Channel	Mid Channel	Rt Channel			Coarse	Medium	Fine	Coarse	Medium	Fine	
1	66.06	666.73	880.57	1613.36	0.200	0.187	0.217	0.201	1.142	10.772	16.510	28.018	13.939	–	–	–	–	–	–	
2	53.40	663.24	673.24	1389.86	0.237	0.343	0.223	0.268	1.093	19.655	12.971	32.183	12.008	0.291	8.479	91.230	1.390	4.447	4.835	
3	64.85	696.60	678.42	1439.87	0.178	0.143	0.228	0.183	0.997	8.607	13.364	22.766	12.440	–	–	–	–	–	–	
4	–	–	–	–	–	–	–	–	–	–	–	–	–	–	–	–	–	–	–	Sunday
5	82.60	545.19	719.94	1347.51	0.127	0.076	0.127	0.104	3.674	10.948	14.786	12.108	11.642	–	–	–	–	–	–	
6	48.25	631.57	627.16	1306.98	0.157	0.100	0.210	0.156	0.655	5.457	11.379	17.615	11.292	–	–	–	–	–	–	
7	94.25	648.44	638.88	1381.57	0.110	0.120	0.075	0.102	0.896	6.723	4.140	12.176	11.937	–	–	–	–	–	–	
8	50.43	636.80	774.40	1461.63	0.200	0.170	0.145	0.172	0.871	9.353	9.702	21.721	12.628	–	–	–	–	–	–	
9	62.74	623.07	715.40	1401.21	0.260	0.210	0.205	0.225	1.409	11.305	12.671	27.240	12.106	–	–	–	–	–	–	
10	65.38	626.62	670.84	1363.04	0.115	0.176	0.153	0.149	0.652	9.637	8.868	17.547	11.777	1.680	1.680	26.620	0.409	0.409	23.444	
11	–	–	–	–	–	–	–	–	–	–	–	–	–	–	–	–	–	–	–	Sunday
12	38.84	525.94	585.28	1210.06	0.160	0.095	0.165	0.140	1.366	4.317	8.343	14.637	10.455	–	–	–	–	–	–	
13	69.35	622.49	630.22	1322.06	0.100	0.150	0.143	0.151	0.599	8.067	7.786	17.248	11.423	–	–	–	–	–	–	
14	61.13	535.48	638.67	1234.98	0.090	0.065	0.140	0.098	0.475	3.007	7.725	10.457	10.670	–	–	–	–	–	–	
15	70.95	746.47	644.83	1462.25	0.210	0.150	0.140	0.167	1.287	9.674	7.800	21.099	12.634	–	–	–	–	–	–	
16	71.69	654.44	628.00	1354.13	0.078	0.113	0.210	0.134	0.483	6.389	11.394	15.676	11.699	–	–	–	–	–	–	
17	61.45	765.62	575.92	1402.99	0.180	0.225	0.123	0.176	0.955	14.883	0.120	21.334	12.122	2.220	8.880	88.880	0.704	2.806	28.059	
18	–	–	–	–	–	–	–	–	–	–	–	–	–	–	–	–	–	–	–	Sunday
19	94.78	713.32	639.90	1448.02	0.220	0.185	0.200	0.202	1.801	11.402	11.057	25.272	12.511	–	–	–	–	–	–	
20	70.80	699.54	532.54	1302.88	0.155	0.150	0.170	0.158	0.948	9.066	7.822	17.786	11.257	–	–	–	–	–	–	
21	87.52	580.98	607.67	1276.17	0.135	0.145	0.155	0.145	1.021	7.279	8.100	15.967	11.026	–	–	–	–	–	–	
22	44.47	692.75	591.78	1329.00	0.170	0.175	0.185	0.177	0.653	10.474	9.459	20.323	11.482	–	–	–	–	–	–	
23	46.38	470.33	829.33	1346.04	0.115	0.166	0.175	0.152	0.461	6.745	12.539	17.676	11.629	–	–	–	–	–	–	
24	76.96	573.56	685.56	1336.08	0.170	0.140	0.185	0.165	1.130	6.938	10.958	19.047	11.544	1.800	7.220	90.960	0.404	1.615	20.325	
25	–	–	–	–	–	–	–	–	–	–	–	–	–	–	–	–	–	–	–	Sunday
26	112.87	575.62	739.75	1428.31	0.200	0.168	0.165	0.177	1.950	8.355	1.950	21.842	12.340	–	–	–	–	–	–	
27	82.14	628.26	583.75	1294.15	0.120	0.130	0.220	0.157	0.852	7.057	11.108	17.554	11.181	–	–	–	–	–	–	
28	72.62	519.42	642.34	1234.38	0.153	0.090	0.205	0.176	0.960	7.629	11.377	18.770	10.665	–	–	–	–	–	–	
29	66.86	517.00	592.95	1176.81	0.140	0.090	0.170	0.133	0.809	4.020	8.709	13.522	10.167	–	–	–	–	–	–	
30	91.31	472.08	580.37	1143.76	0.454	0.074	0.104	0.210	3.582	3.018	5.215	20.752	9.882	–	–	–	–	–	–	
31	24.30	466.09	675.65	1166.04	0.140	0.166	0.375	0.215	0.218	6.685	20.891	21.659	10.074	9.030	2.400	88.550	1.749	0.468	17.140	

66

5.5   BED LEVEL CHANGES

The changes that occur in bed level of the river as a result of the changes in tidal flow characteristics can be related to sediment transport rate. The continuity equation of sediment mass can be expressed as vide eqn. (5.14):

$$\frac{\partial Q_{bx}}{\partial x} + \frac{\partial Q_{by}}{\partial y} + \gamma_b\,(1-\xi)\,\frac{\partial z_0}{\partial x} = E - D \qquad \dots\ (5.14)$$

where Q_{bx}, Q_{by} represent bed load transport rate in wt/unit width in the longitudinal and transverse directions ξ is porosity of the bed material; γ_b the specific weight of the sediment and Z_0 the elevation of the sediment surface above a horizontal datum.

E is the quantity of sediment removed into the suspension from the moving bed layer in wt/unit area of bed surface per unit time and D is the quantity of sediment deposited into the mobile bed in wt/unit area/unit time.

The distribution of suspended material in an environment can be determined from the sediment mass continuity equation:

$$\frac{\partial C}{\partial t} + \bar{u}\frac{\partial C}{\partial x} + \bar{v}\frac{\partial C}{\partial y} + (w - \omega_f)\frac{\partial C}{\partial z} = \frac{\partial}{\partial_y}\left(\varepsilon_y\frac{\partial C}{\partial y}\right) + \frac{\partial}{\partial x}\left(\varepsilon_z\frac{\partial C}{\partial z}\right) \dots (5.15)$$

where C is the turbulent average concentration of sediment in mass/unit volume of sediment, water mixture; ε_y, ε_z are the lateral and vertical diffusion coefficients and ω_f the fall velocity. The changes in elevation of the bed surface can be predicted provided sediment discharge ratios and sediment pick up and deposited parameters are known as functions of space and time. Unfortunately, the precise form of the parameters is unknown in tidal environment and engineering problems involving bed sediment transport in general have to be solved by field observations. Integration of eqns (5.14) and (5.15) over the channel cross-section and with time indicate that the stability condition of the channel can be expressed as

$$\int_{t_1}^{t_2}\frac{\partial Q_{Tx}}{\partial x}\,dt + \Delta Q_{Ty} + \Delta C_t = 0 \qquad \dots\ (5.16)$$

where x, y are the co-ordinates along and normal to the channel axis and Q_{Tx} total transport rate over the channel cross-sections. ΔQ_{Ty} is the net lateral transport rate per unit channel length in time interval $t = (t_2 - t_1)$ and ΔC_t is the change in cross-sectional average suspended concentration over time.

All the quantities in the equation can be determined from the analysis of measured field data along with suitable sediment transport formula, which will provide a means to assess whether the channel is silting or scouring.

Sedimentation in a navigational channel can be assessed from the difference in total load between the flood and ebb cycle. Total load consists of both suspended and bed load. The suspended load may be predominantly fine sand, i.e., non-cohesive or clay or clay and silt, i.e., cohesive flocculent type, depending on the composition of the bed material load. Estimation of suspended load is normally made from the measured suspended load concentration and velocity over a tidal cycle at the measuring station. The bed load cannot be measured and has to be calculated or alternately assumed to be a certain fraction of the suspended load. The bed shear stress over the tidal cycle can be obtained with the assumption of the existence of steady flow velocity profile close to the hydraulically rough boundary.

The bed load transport rate can then be ascertained using suitable formula. The total sediment flux can be obtained after ascertaining the suspended load component, which equals the depth-integrated product of current velocity and suspended material concentrations.

Estimation of bed load can be made by use of any suitable bed load formula. One such formula is that of Einstein and Brown [5]. It can be expressed as:

$$\phi = 40 \left(\frac{1}{\psi} \right)^3 \qquad \dots (5.17)$$

$$\phi = \frac{q_{bv}}{\gamma_s K \sqrt{g \dfrac{\gamma_s}{\gamma} - 1}} \, d_{50}^3 \qquad \dots (5.18)$$

$$\frac{1}{\psi} = \frac{\tau}{(\gamma_s - \gamma) d_{50}} \qquad \dots (5.19)$$

$$K = \sqrt{\frac{2}{3} + \frac{36 \, v^2 \gamma}{g \, d_{50}(\gamma_s - \gamma)}} - \sqrt{\frac{36 \, v^2 \gamma}{g \, d_{50}(\gamma_s - \gamma)}} \qquad \dots (5.20)$$

On the other hand, for estimation of the total sediment load the Engelund-Hansen [6] formula can be adopted. The formula is

$$q_{tc} = \frac{0.05 \, V^5}{(s - 1)^2 \, g^{1/2} \, d_{50} \, C_c^3} \qquad \dots (5.21)$$

where g is acceleration due to gravity; d_{50}–representative grain size diameter; C_c–Chezy's coefficient; v–kinematic viscosity of water; q_{bv}–bed load transport; γ_s–sp.wt of water and sediment particle, q_{tc}–total bed material transport; V–depth averaged velocity and s–specific gravity of the sediment.

5.6 PROCESSING OF DATA

Hydrographic data measurements primarily include velocity, both magnitude and direction, sediment concentration, conductivity or salinity and temperature at several depths between the surface and the bottom at a number of different stations. In an estuary of complex shape and bathymetry, velocity measurements are made for both the current speed and direction. The velocity output then consists of a longitudinal velocity component considered negative in the main ebb direction and lateral component negative 90° clockwise relative to the positive longitudinal direction. Here it is necessary to specify the ebb orientation of the estuarine location relative to the north. Along with current speed, data of water samples are also collected for conductivity or salinity and temperature measurements. Samples at different locations along the vertical are also taken for laboratory analysis of suspended load concentration.

To find out the net sediment flux it is convenient to compute time average or net value in terms of non-dimensionalised depths varying from 0 to 1, i.e., bottom to surface rather than values collected at a fixed distance below the surface or above the bottom. Measurements are made generally at a constant sampling interval of t over at least one complete tidal cycle.

Further, it is expected that measurements begin at slack water, i.e., close to either high or low tide. For a semi-diurnal tide, if the number of intervals selected is 12, the sampling rate then approximates one lunar hour. Generally the period is seldom an exact multiple of the sampling rate. It is necessary to plot curves of each variable at all non-dimensionalised depths as funtion of time and then divide each time series into n equal increments. The interpolated values at each n at the $(n + 1)$ times would then be the data to be used in computing time averages. The net value of the various quantities based on the measured values such as discharge, sediment fluxes, can then be estimated based on the relations discussed in tidal analysis.

The sedimentation rate can be assessed from the difference of total load between the flood and the ebb cycle. Processing of data can be carried out with the development of a suitable software program. The flow chart of the developed software is shown, in Appendix II.

5.7 CASE STUDY

The software (AHDTMP) was applied to process the measured data of Hooghly River. Figure 5.2 shows the time variation of velocity, sediment concentration and water level during a tidal cycle. Based on this input the vertical profile for velocity (Fig. 5.5), suspended sediment (Fig. 5.6) was built up. The information was then adopted to obtain the variation of bed shear stress over the tidal cycle (Fig. 5.7).

The relation between bed shear stress and current velocity can be expressed in terms of a non-dimensionalised drag coefficient as

$$C_1 = \left(\frac{u_*}{u_1} \right)^2 \qquad \qquad \dots (5.22)$$

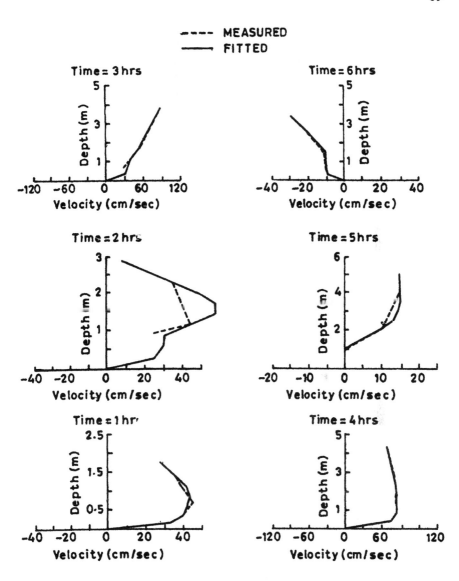

Fig. 5.5 Velocity profiles

where u_1 is the current speed 1 m above the bottom. For a given boundary surface C_1 approaches a constant value whenever the appropriate Reynolds number exceeds 10^5 to 1.5×10^9 which corresponds to u_1 value in the range of 10–15 cm/s [9].

70

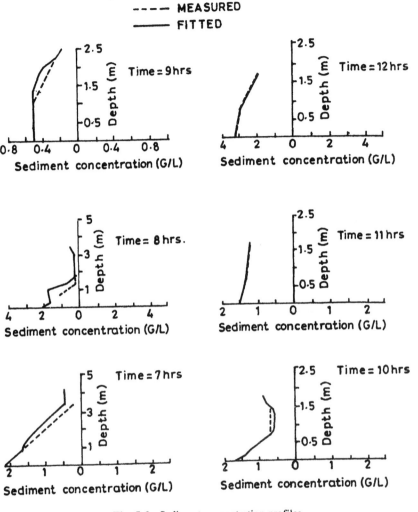

Fig. 5.6 Sediment concentration profiles

For most estuarine applications it seems reasonable to set $C_1 = 0.003$. For this value of C_1 and the measured u_1 it is possible to solve for the friction velocity u_*. Knowing u_* the boundary shear stress can be obtained from the relation

$$u_* = \sqrt{\frac{\tau_0}{\rho}}$$

.... (5.23)

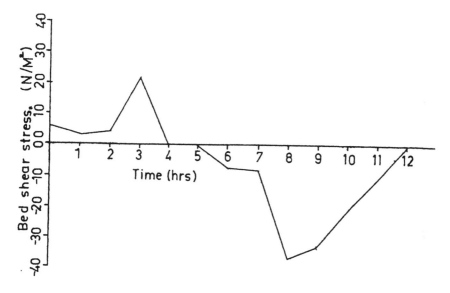

Fig. 5.7 Variation of bed shear stress during a tidal cycle

The value of $C_1 = 0.003$ implies a fixed value of z representing the dynamic roughness length, which means the distance above the bottom where the velocity is zero in the logarithmic velocity law. Although z is variable it does not have as large an effect on the drag coefficient. The steady state of flow assumption necessary to assume the existence of the logarithmic current speed profile is not restrictive. The current speed varies primarily on a time scale of the tidal period whereas u_1 is a short-term mean on the order of 1 min or so.

Having obtained the bed shear stress the bed load transport rate following the Einstein-Brown relation can be obtained. In the present case study the representative grain size d_{50} has been considered to be fine sand equal to 0.125 mm. Accordingly, the net bed load and net suspended load transport over the tidal cycle has been evaluated in volumetric rate. The total transport rate has also been obtained using the Engelund-Hansen [6] formula over the tidal cycle. Table 5.2 shows the results of computation:

Table 5.2 Results of computation of sediment flux

Suspended sediment flux:	– 6.63 m³/m width
Bed load:	– 0.02 m³/m width
Total load transport :	– 6.65 m³/m width
Total load transport (Engelund-Hansen):	– 10.65 m³/m width
(Negative sign indicates net transport in the direction of ebb)	

72

LITERATURE CITED

1. Dillo. H.G. (1960). Sandwanderung in Tideflussen. Mitteilundgen des Franzius-Institute for Grundund Wasserbau der Technischen Hochschule, Hannover.
2. Kestner. F.J.T. (1961). Short term changes in the distribution of fine sediments in estuaries. *Proc. Inst. Civil Engineers*, 185–208.
3. Biswas, A.N and Mitra, B. (1979). Assessment of reshoaling in a dredged cut in a tidal alluvial waterway. *Irrigation and Power*.
4. Vanoni, V.A. (1975). *Sedimentation Engineering*. Amer. Soc. Civil Engineers,
5. Brown, C.B. (1950). *Engineering Hydraulics*. (H. Rouse, ed.) John Wiley and Sons, New York.
6. Engelund, F. and Hansen. E. (1967). A monograph on sediment transport in alluvial stream. *Nordic Hydrology*. No. 7.
7. Dyer. K.R. (1973). *Estuaries: A Physical Introduction*. John Wiley and Sons, New York.
8. Morra, R.H.J. (1948). The sand concentration vertical for steady flow under equilibrium conditions. Report No. 9. Rijkswaterstart, directie Bendenriviern. (in Dutch).
9. Sternberg, (1968). Friction factors in tidal channels with differing bed roughness, Marine Geology 6: 243–260.

APPENDIX I: COMPUTER PROGRAM

```
*          dimension  d (20), v(20,20), w(20,20), th(20,20), b(20,20), a(200)
           open (54,file = 'TDL. IN')
           open (88, file = 'TDL. OUT')
*
*          write (*,*) 'give the input data'
*
           read(*,*) t, nt
   1       if (t-999)3,500,500
   3       do 200 i = 1, nt
 200       read (54,*)  d (i), (v (i, j) ,w (i, j), th (i, j), j = 1,6)
           do 251 i =1, nt
           do 252 j =1,6
           if (i. eq. 1. or. i. gt. 9) then
           th (i,j) = abs (th (i, j ) –273)
           if (th (i, j). ge.180) then
           th (i,j) = abs (th (i, j)–180)
               else
           end if
               go to 252
           else
           th (i, j) = abs (th (i, j)–93)
           end if
 252       continue
 251       continue
           write (88, 89) ((i, j, th (i, j), j = 1,6), i = 1, nt)
  89       format ((2x, i2, 3x, 12, 5x, f6.2)/)
           do 250 i = 1, nt
           do 250 j = 1, 6
 250       b(i, j) = v (i, j) * w (i, j) * cos (th (i,j))
           a (1) = 0.0
           do 10 l = 1, nt
           do 11 j = 1,6
           write (*,*) a (i)
  11       a (j+1) = a(j) + ( (b (i, 1) + (b (i,6))/2) * 0.2 * (j–1) * d (i))
           a(i) = a(j)
  10       continue
           i = 1
           s1 = 0
           s2 = 0
           if (a  (i)) 25, 30, 30
  30       if (a (i + 1)) 35, 40, 40
  40       s2 = s2 + (a(i) + a (I + 1))*.05
           i = i + 1
           if (i – nt) 35, 50, 35
  35       x = a (i) / (a(i) – a (i + 1))
           s2 = s2 + a (i) *x*.5
           s1 = s1 + a (i) * (1–x) *.5
```

```
        i = i + 1
        if ( i – nt) 25, 50, 25
25      if (a (i + 1) ) 26, 31, 31
26      s1 = s1 + (a (i)  + a (i + 1)) *.5
        i = i + 1
        if (i – nt) 25, 50, 25
31      x = a (i) / (a (i + 1) – a (i))
        s1 = s1 + a (i) * x  * . 5
        s2 =  s2 + ( 1 – x ) *a ( I + 1 ) * .5
        i = i + 1
        If  ( i – nt) 30, 50, 30
50      if (a (nt – 1)) 65, 70, 70
70      if (a (nt)) 75, 80, 80
80      a (nt) = a (nt – 1) + (a (nt) – a (nt – 1) ) * t
        s2 = s2 + (a (nt – 1) + a (nt)) * . 5 * t
        go to 150
75      y = a (nt – 1) / (a(nt – 1 ) – a (nt))
        if (t – nt) 90, 90, 85
90      a (nt) = a (nt – 1) –a (nt-1) * t/y
        s2 = s2 + ( a (nt – 1 ) + a (nt))* . 5*t
        go to 150
85      a (nt) = (t – nt) * (a (nt) / (1 – y ))
        s1 = s1 + 0.5 * a (nt) * (t – y)
        s2 = s2 + 0. 5 * y * a (nt – 1)
        go to 150
65      if (a (nt)) 100, 110, 110
100     a (nt) = a (nt – 1) + (a (nt ) – a (nt – 1) ) * t
        s1 = s1 + (a (nt – 1) + a (nt)* .5*t
        go to 150
110     y = a (nt – 1 )/(a (nt) – a (nt – 1))
        if (t – y) 120, 120, 115
120     a (nt ) =  (a (nt – 1) –a (nt – 1)) * t/y
        s1 = s1 + (a (nt – 1) + a (nt) ) *.5 * t
        go to 150
115     a (nt) = (t – y) * a (nt) / (1 – y)
        s1 = s1 + 0.5 * y * a (nt – 1)
        s2 = s2 + 0.5 *a (nt) * (t – y)
        go to 150
150     s1 = –s1 * 3.6
        s2 = s2 * 3.6
        r = s1/ (s1 + s2)
        do 551 i = 1, nt
        write (88, 93) d(i)
        do 552 j = 1, 6
552     write (88, 43) v (i, j), w (i, j) , th (i, j)
551     continue
        write (*, 42)
        write ( 88, 22) sl, s2, r
        go to 1
42      format (20(/) )
```

```
43      format ( 3(4x, f5, 1) / )
93      format ('DEPTH IN MTS', f 5.1)
22      format (2x, ' EBB DISCHARGE ', f20. 2//2x, 'FLOOD DISCHARGE',
        'f20.2//2x,' 'RATIO', f10.8/)
500     stop
        end
```

1	1	63.00
1	2	33.00
1	3	30.00
1	4	48.00
1	5	93.00
1	6	93.00
2	1	93.00
2	2	93.00
2	3	93.00
2	4	93.00
2	5	56.00
2	6	56.00
3	1	10.00
3	2	8.00
3	3	2.00
3	4	3.00
3	5	3.00
3	6	2.00
4	1	6.00
4	2	7.00
4	3	10.00
4	4	13.00
4	5	9.00
4	6	10.00
5	1	7.00
5	2	2.00
5	3	10.00
5	4	9.00
5	5	9.00
5	6	2.00
6	1	5.00
6	2	5.00
6	3	8.00
6	4	10.00
6	5	10.00
6	6	4.00
7	1	10.00
7	2	4.00
7	3	4.00
7	4	0.00
7	5	1.00

7	6	3.00		
8	1	1.00		
8	2	5.00		
8	3	3.00		
8	4	8.00		
8	5	2.00		
8	6	3.00		
9	1	1.00		
9	2	22.00		
9	3	2.00		
9	4	80.00		
9	5	93.00		
9	6	93.00		
10	1	92.00		
10	2	15.00		
10	3	36.00		
10	4	36.00		
10	5	14.00		
10	6	22.00		
11	1	17.00		
11	2	17.00		
11	3	16.00		
11	4	23.00		
11	5	28.00		
11	6	26.00		
12	1	17.00		
12	2	32.00		
12	3	19.00		
12	4	16.00		
12	5	16.00		
12	6	30.00		
13	1	28.00		
13	2	26.00		
13	3	31.00		
13	4	12.00		
13	5	22.00		
13	6	43.00		
14	1	39.00		
14	2	35.00		
14	3	33.00		
14	4	28.00		
14	5	25.00		
14	6	32.00		
	DEPTH IN MTS 8.2			
		21.0	0.6	63.0
		21.0	0.8	33.0
		21.0	0.8	30.0
		10.0	1.4	48.0
		0.0	1.1	93.0
		0.0	1.1	93.0

DEPTH IN MTS 7.3

0.0	0.3	93.0
0.0	0.3	93.0
0.0	1.1	93.0
0.0	1.3	93.0
5.0	1.4	56.0
8.0	2.7	56.0

DEPTH IN MTS 6,7

67.0	0.8	10.0
62,0	1.0	8.0
56.0	1.1	2.0
62.0	1.4	3.0
56.0	2.0	3.0
51.0	2.1	2.0

DEPTH IN MTS 6.4

82.0	0.2	6.0
87.0	0.4	7.0
87.0	0.6	10.0
82.0	0.6	13.0
77.0	0.6	9.0
63.0	1.0	10.0

DEPTH IN MTS 6.1

92.0	0.7	7.0
87.0	1.9	2.0
79.0	1.3	10.0
72.0	1.5	9.0
56.0	1.4	9.0
54.0	0.0	2.0

DEPTH IN MTS 5.5

90.0	0.9	5.0
87.0	1.7	5.0
85.0	1.2	8.0
77.0	0.0	10.0
62.0	1.5	10.0
51.0	1.5	4.0

DEPTH IN MTS 5.2

77.0	0.8	10.0
74.0	1.5	4.0
72.0	1.4	4.0
69.0	1.4	0.0
46.0	0.0	1.0
46.0	1.4	3.0

DEPTH IN MTS 4.9

62.0	1.0	1.0
56.0	0.8	5.0
56.0	1.6	3.0
56.0	1.5	8.0

78

```
            46.0   1.4    2.0
            36.0   1.7    3.0
```

DEPTH IN MTS 4.4
```
            31.0   1.1    1.0
            10.0   1.1   22.0
             5.0   1.2    2.0
             5.0   1.5   80.0
             0.0   0.0   93.0
             0.0   1.3   93.0
```

DEPTH IN MTS 5.2
```
            31.0  290.0  92.6
            31.0   0.3   15.0
            33.0   0.9   36.0
            33.0   1.3   36.0
            38.0   1.4   14.0
            31.0   1.3   22.0
```

DEPTH IN MTS 6.6
```
            67.0   0.7   17.0
            62.0   0.0   17.0
            67.0   0.9   16.0
            62.0   0.9   23.0
            56.0   1.0   28.0
            41.0   1.1   26.0
```

DEPTH IN MTS 7.6
```
            77.0   0.0   17.0
            62.0   0.0   32.0
            52.0   0.8   19.0
            46.0   0.9   16.0
            41.0   0.9   16.0
            36.0   1.1   30.0
```

DEPTH IN MTS 8.1
```
            67.0   0.7   28.0
            56.0   0.2   26.0
            56.0   0.8   31.0
            46.0   1.0   12.0
            41.0   0.9   22.0
            13.0   1.1   43.0
```

DEPTH IN MTS 8.5
```
            41.0   0.4   39.0
            38.0   0.9   35.0
            30.0   0.9   33.0
            31.0   1.0   28.0
            23.0   2.3   25.0
             8.0   1.2   32.0
```

EBB DISCHARGE   − 72160.28
FLOOD DISCHARGE   310244.0
RATIO   − .30308690

APPENDIX II: FLOW CHART FOR AHDTMP

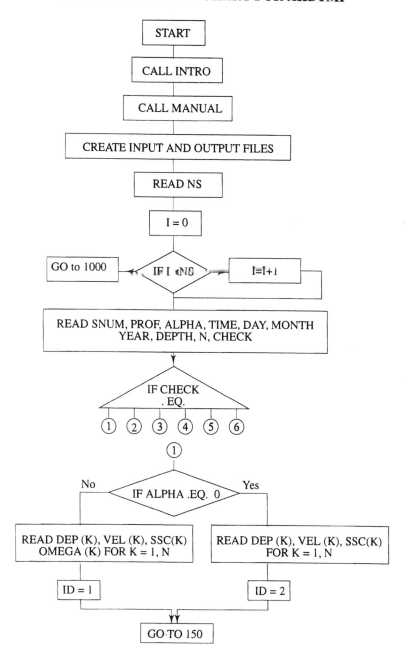

80

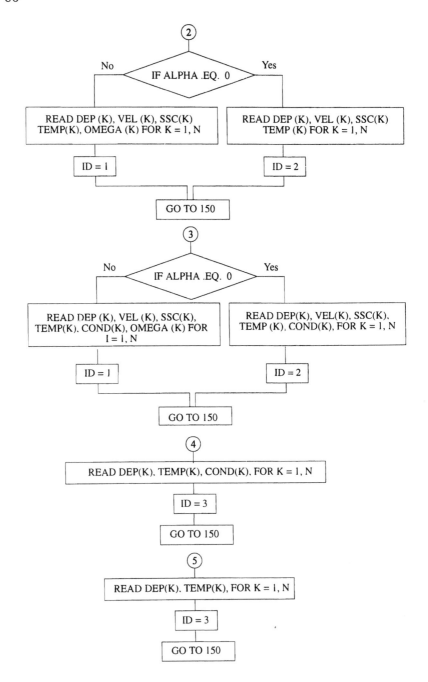

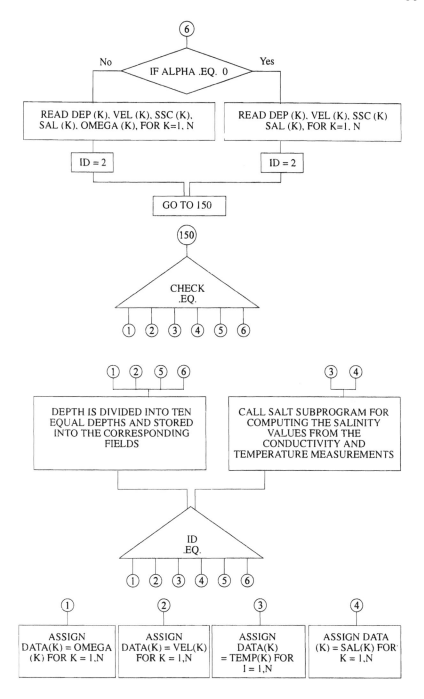

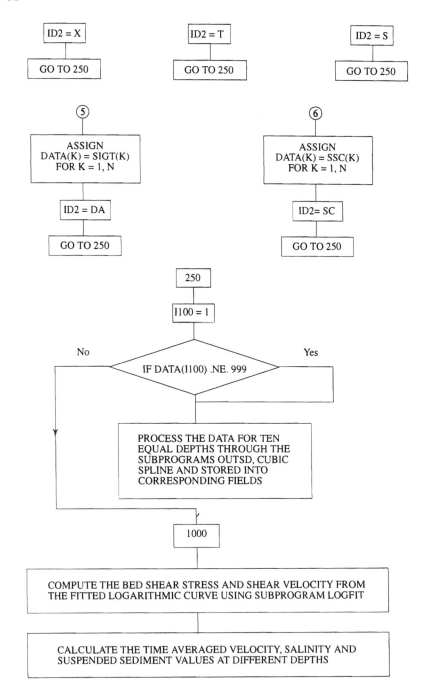

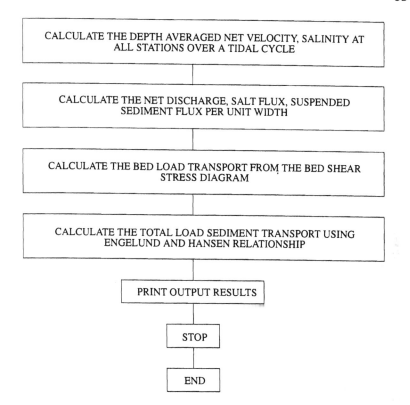

CALCULATE THE DEPTH AVERAGED NET VELOCITY, SALINITY AT ALL STATIONS OVER A TIDAL CYCLE

CALCULATE THE NET DISCHARGE, SALT FLUX, SUSPENDED SEDIMENT FLUX PER UNIT WIDTH

CALCULATE THE BED LOAD TRANSPORT FROM THE BED SHEAR STRESS DIAGRAM

CALCULATE THE TOTAL LOAD SEDIMENT TRANSPORT USING ENGELUND AND HANSEN RELATIONSHIP

PRINT OUTPUT RESULTS

STOP

END

CHAPTER 6

HYDROGRAPHIC SURVEYS IN TIDAL RIVERS

6.1 INTRODUCTION

A hydrographic survey usually consists of determination of the following types of data: (i) topography of the bottom, (ii) heights and times of tides and water stages and (iii) location of fixed objects for survey and navigational purposes. In general a hydrographic survey means the procedure of measuring the depth of water in a river or sea utilising a boat or survey ship as a working platform while simultaneously determining the horizontal position of the boat relative to the near-by shoreline and prominent natural or man-made objects.

Data for each sounding made must include the depth of water, the horizontal position of the place at which the depth measurement was made and the elevation of the water surface. The elevation of the water surface is usually made by an automated recording instrument at a place other than that of the depth measurement and connected to it by noting the time of measurement. The water surface record is necessary since the depth at any location varies depending on the tidal, meteorological or other man-made effects. These effects in most cases can be determined from the variations of height of the water at some reference station such as at a recording tide station.

6.2 CONTROL

It is necessary to establish adequate horizontal and vertical control in and surrounding the hydrographic survey area. The reference point or datum used for control of the surveys should be recoverable. This can be accomplished by tying to an existing geodetic control network, preferably the national geodetic network. In most situations the existing horizontal network will not provide sufficient coverage for the entire area and will need to be extended by traverse, triangulation, utilising theodolite electronic distance measurement.

The vertical datums used may be the local tidal or engineering datums. The datums used should be permanent and well-defined in nature. Since all surroundings are made relative to the water surface at the time of the sounding it is necessary to reference the water surface to some permanent datums at each instant of time for the entire survey.

Normally the national survey organisation, such as the Survey of India, will establish a system of bench-marks at each of its tide observation stations. They provide permanent reference points for tide level observations and for the datum planes determined from these observations.

The shoreline is also required to be mapped with the same control as exercised for the hydrographic survey. This can be achieved by using a plane table or similar topographic method. The shoreline may also be charted at high tide using a boat and hydrographic survey procedure. The reference level may be the M.L.W. or M.H.W.

6.3 POSITIONING SYSTEM

A number of methods are available for determining the horizontal position of the craft at the time of sounding. All systems can basically be classified as either visual or electronic survey methods. In the visual approach the methods are as follows.

6.4 VISUAL

6.4.1 Direct

Herein the distance is measured directly between a reference point on the shore and the sounding point. A distance line is used which is stretched between the vessel and the bank from which the distance can be read between the reference mark and the sounding point.

The method is mostly suited for use along banks and other shallow waters where sufficient draught is not available for a survey launch.

6.5 OPTICAL INSTRUMENT METHOD

6.5.1 Single Horizontal Sextant Angle Subtense (Fig. 6.1)

Herein the position of any point on the axis of the cross-section can be computed from the observed subtense angle α. In case a fixed subtense base AC is used the subtense angle α belonging to the distance OA of each of the predetermined sounding points O can be computed in advance, thereby easing the plotting process and increasing the speed of operation. This is suitable for surveys up to 200 m from the shore-marks.

86

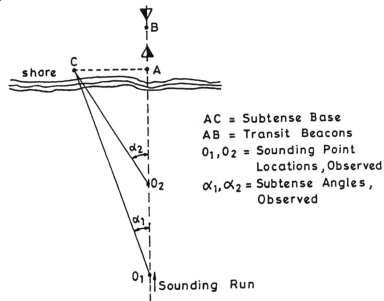

AC = Subtense Base
AB = Transit Beacons
O_1, O_2 = Sounding Point
        Locations, Observed
α_1, α_2 = Subtense Angles,
        Observed

Fig. 6.1 Horizontal sextant angle subtense (single)

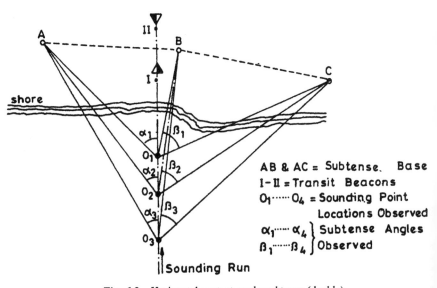

AB & AC = Subtense Base
I - II = Transit Beacons
$O_1 \cdots O_4$ = Sounding Point
        Locations Observed
$\alpha_1 \cdots \alpha_4$ } Subtense Angles
$\beta_1 \cdots \beta_4$ } Observed

Fig. 6.2 Horizontal sextant angle subtense (double)

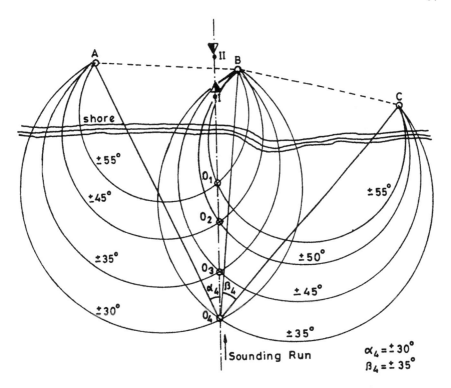

Fig. 6.3   Lattice chart of sextant angles subtense

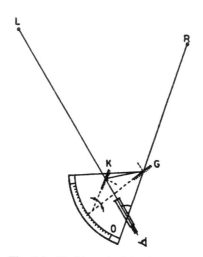

Fig. 6.4   Working principle of a sextant

6.5.2 Method of Resection or Double Horizontal Sextant Angle Subtense

Herein a three-point fix is obtained using two sextants on the vessel to measure the two adjacent angles between three signals at known locations on shore. Each of the angles α and β (Fig. 6.2) has its specific subtense base line AB and BC respectively. The beacons I and II represent the line of the cross-section which will help the captain of the vessel to maintain course in line with the section. Sometimes a sextant lattice chart which consists of arcs can be constructed in advance (Fig. 6.3).

This simplifies plotting of the sounding points later. Figures 6.4 and 6.5 show a drawing of a sextant and *using it the angles to be observed shall not be smaller than 20° and also shall not be above* 120° so that the accuracy required is maintained.

6.6 ELECTRONIC POSITIONING SYSTEM

This usually consists of a distance measuring system using a radio transmitter erected at each of two or more known shore locations and ship-bound receiver. Range-range positioning systems basically consist of three units— two shore receiver transmitter units at known locations and one shipboard unit which activates the shore transmitters by pulse modulation or phase change. The resultant signal received at the ship is used to directly calculate the distance to each of the two shore locations. By trilateration the position of the survey vessel can be computed.

6.6.1 Hi-Fix System

This is a precision lightweight electronic position fixing system with 24-volt dc operation developed by the Decca Navigator Company. The system comprises a master station and two slave stations. The slave stations are shore stations, fixed in position, while the master station may be either onshore or on the survey vessel. In the former any number of vessels with a Hi-Fix receiver onboard can use the system, whereas in the latter only one vessel carrying the master station can use it. In this system the phases of the two separate sets of electromagnetic waves are compared. One set comprises a phase comparison of waves generated by the master and one slave station and the other set a phase comparison of waves generated by the master and the second slave station (Fig. 6.6).

At an instant t the electromagnetic field radiated by both the stations A and B may be represented respectively by

$$h_A = H_A \sin 2\pi ft \qquad \dots \dots (6.1)$$

$$h_B = H_B \sin 2\pi ft \qquad \dots \dots (6.2)$$

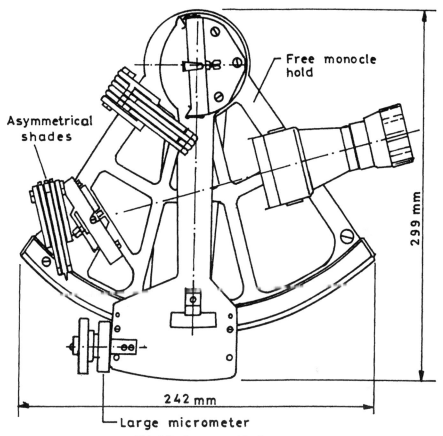

Free monocle hold

Asymmetrical shades

299 mm

242 mm

Large micrometer

Fig. 6.5 A sextant with all components

f being the frequency of the transmitted wave. The fields prevailing at an instant t at Q produced by radiations from stations A and B will be

$$h_{AQ} = g_A H_A \sin 2\pi f \left(t - \frac{d_A}{V} \right) \qquad \text{.... (6.3)}$$

$$h_{BQ} = g_B H_B \sin 2\pi f \left(t - \frac{d_B}{V} \right) \qquad \text{.... (6.4)}$$

where V is the velocity of propagation. Their phase difference

$$\phi = \frac{2\pi f}{V} (d_B - d_A) \pm 2K\pi = \frac{2\pi}{\lambda} (d_B - d_A) \pm 2K\pi \quad \text{.... (6.5)}$$

where K is an integer and λ is the wavelength:

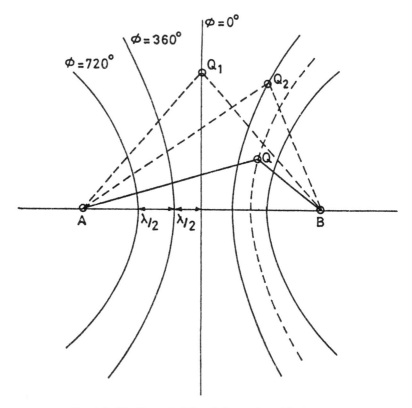

Fig. 6.6 Working principles of electronic positioning system

$$d_B - d_A = \frac{\phi\lambda}{2\pi} \pm K\lambda \qquad\qquad \dots (6.6)$$

which shows that the locus of points where the phase difference is constant is a hyperbola. When $\phi = 0$, one gets, $d_B - d_A = \pm K\lambda$, an expression defining a family of homofocal hyperbolas whose foci are located at stations A and B. K is assigned a positive or negative value depending on whether the equiphase hyperbola is to the right or left of the bisector which is expressed in degrees. Again, between station A and another station C, a second set of homofocal hyperbolas can be developed (Fig. 6.7). The position of the receiving antenna (SV_1 and SV_2 etc.) defines a point which is the intersection of one hyperbola of the first set with that of another belonging to the second set. However, when any position A is shifted to the observer's point Q, the system immediately reduces to a set of concentric circles and hence the second type of Hi-Fix operation of the two-range system follows (Fig. 6.8). Along with the receiver on the vessel, a track plotter can be used which

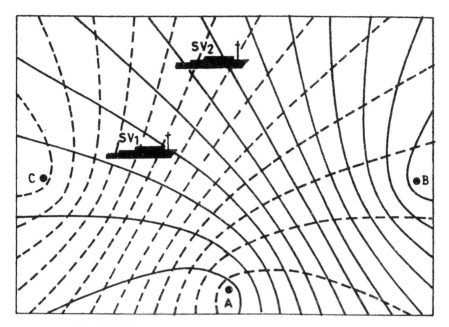

Fig. 6.7 Electronic positioning system hyperbolic mode (typical)

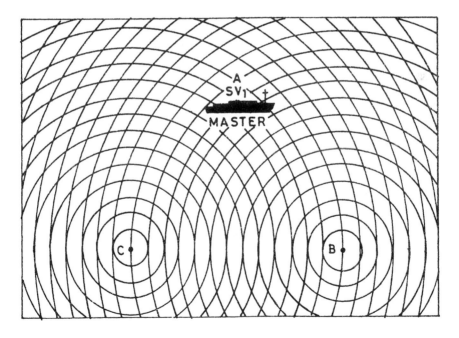

Fig. 6.8 Electronic positioning system circular mode (typical)

gives a pictorial representation of the working area. A chart moves the rollers and stylus which moves perpendicular to the movement of the chart and shows the position as well as the course of the vessel.

Hydrographic positioning systems are passive systems, that is the shipboard unit consists of a receiver only. Such systems require a minimum of three shore transmitting stations of known position. Because the system is passive an unlimited number of vessels can use this hyperbolic system simultaneously.

The various patented systems for two-dimensional positioning systems are Motorola, Decca which operates on circular modes and Syledis which operates in the hyperbolic mode.

They may operate with regard to the (i) frequency of the electromagnetic waves or (ii) the difference in the method of distance measured between the master and the remote station. In other systems the time taken by the singal to travel vice versa between the master and the slave is measured.

The circular system uses short range microwave systems with a frequency range between 5 to 9, 10^3 Hz with omni-directional antenna and one master on board and 2 to 4 remotes on shore. Here the distance is measured by the time interval method. In the hyperbolic system the distance is measured by phase difference with a frequency range between 2 to 3, 10^3 Hz with remote stations having fixed positions.

All modern systems are fitted with binary-coded outputs for processing via computer and automatic printout or plotting of the fixes at required intervals. A track plotter with additional left-right indicator can be incorporated to enable the captain of the survey launch to keep the launch on course along a predetermined track.

Electronic positioning systems in use require field calibration. This is achieved by comparing the position obtained by the visual method or by the more accurate electronic system. The accuracy of electronic systems depends on many factors, one of which is the performance of the equipment. Factors such as propagation anomalies, reflections both from sea and vertical surfaces, shadowing by islands and geometry between the lines of position are all important to the precision with which the equipment can observe a range.

6.7 SOUNDING

Sounding is a measurement of the depth of water utilising an echo-sounder or a lead line. Echo-sounding is the method of measuring the depth of water by determining the time required for sound waves to travel at a known velocity from a point near the surface of the water to the bottom and return. Echo-sounding equipment is designed to produce the sound, receive and amplify the echo, measure the intervening time interval and convert this interval automatically into units of depth measurements such as fathoms or metres. Some echo-sounders can measure depths as shallow as a few metres or feet while others can measure in excess of several thousand metres. The depth measurements by echo-sounder are essentially continuous so a survey vessel need not stop to take a sounding but can continue in speed. The echo-sounder needs to be calibrated to take care of variations in water density and vessel transducer depth. Calibration may be done on a small vessel by a bar check. This consists of lowering a flat metal bar to various known depths under the survey vessel and calibrating it on these known depths. The bar should be of a length approximately equal to the beam of the boat and suspended on one end from each side of the boat in a horizontal position directly under the sounding transducer. For range vessels working in areas of level bottom a series of lead-line measurements to the bottom taken simultaneously with echo-sounder readings will serve to calibrate the echo-sounder. A lead line usually is a braided cotton rope with a phosphor-bronze wire centre, marked in metres with attached cords, leather straps and coloured bunting and a 2.7 or 6.3 kg lead weight fastened to one end of the lead line. To calibrate an echo-sounder the length of the line used must be verified periodically. For this, comparison with a steel tape should be made with tension on the line equal to the weight of the attached lead in water.

6.7.1 Sounding Lines

Surveys by echo-sounder result in a continuous line of sounding in the direction of travel of the vessel, and the variation in depth along the line is displayed in graphic form. The variation along a line perpendicular to line of sounding is found only by carrying out another sounding. Accordingly, sounding lines are oriented such that the greater depth variations are recorded. In river surveys the sounding lines should be within 45 m or less, being perpendicular to the direction of flow. The spacing between the lines should be so selected as to provide a good representation of the bottom. For a desired spacing between sounding of 15 m and between lines of 30 m the scale of survey should be 1:3000. Generally the scale of the field survey is twice the scale of the published chart. The spacing should be designed so as not to miss sandbars or obstructions of significant size while minimising the number of lines.

A record series of cross-lines may be run approximately perpendicular to the first, their spacings may be up to 10 times that of the first series.

Sounding is subjected to systematic errors and hence corrections are needed. Constants derived from calibration of the horizontal control and echo-sounder can be used for making corrections. Additional corrections need to be applied for the water surface elevation, the draught, settlement and the squat of the survey vessel along with wave action on the vessel.

To determine water surface elevation correction the height of water in the area of the survey relative to the survey vertical datums as a function of time needs to be recorded. The water surface elevation correction and height of water are referred to as the tide correction and height of the tide in the case of tidal rivers, and for rivers and lakes subjected to fluctuations due to variation in upland flow as the stage correction and stage for waters. These two values are normally equal in magnitude but opposite in sign. In rivers where the water heights vary seasonally, a suitable staff gauge graduated in metres is mounted vertically on the side of the piling such that the bottom of the staff extends below the lowest low water and the top extends above the highest high water. The elevation of the staff zero should be referred to a near-by bench-mark.

In tidal rivers or for large rivers requiring water height records at many locations, use of automated recording gauges is desirable. Gauge readings should be compared regularly with a staff reading. The soundings are identified by time; the height of water above datums can then be determined for the time of the sounding and the correction applied to the sounding. The height of the water above datums is subtracted from the raw sounding. The depth of water below the datums is added to the raw sounding.

The draught of the survey vessel and the depth of the echo sounding transducer are related by a finite difference which may be near zero. For the given vessel the change in draught due to loading with fuel supplies or personnel will cause changes in depth of the transducer below the water surface. The transducer depth below the water-line may be measured either directly when the vessel is out of the water in dry dock, by a bar check or indirectly with a lead line at various depths. The draught correction is added to the raw sounding data if the draught was assumed to be zero for the survey. Settlement and squat describe the change in draught and trim of a vessel relative to underlay. A vessel underlay will ride lower in the water when not making way until it begins to plane, in which case it rides higher in the water. This change in draught is termed settlement. Squat is the tendency of the vessel bow to rise and stern to sink as the speed of the vessel increases. The settlement and squat corrections may be either positive or negative but their magnitude usually does not exceed 30 cm for most survey vessels. Wave action may cause sounding to vary considerably. Waves increase in magnitude due to roll and heave of the survey vessel. The sounding taken when the vessel is in trough will show a shallower depth than actually exists. Correction for wave action should be approximated by the crew of the vessel at the time of the survey.

6.8   ECHO SOUNDER

The functioning of the equipment can be described as follows (Fig. 6.9). A short but strong electrical impulse is produced by the signal generator, then amplified and sent to the transducer where it is converted into an acoustic signal which is sent to the river-bed. The signal will be reflected by the river-bed and converted into an electrical impulse, which is then amplified by the transducer. The time lapse between transmission and reception of the signal is measured and converted into the depth of water under the trans-ducer. The depths are shown on a suitable indicator, a recorder, or given by means of a digital output. Because of the high transmission frequency the individually measured depths appear on the recording paper as an almost continuous graph representing the bed profile. It is, however, possi-ble to adjust the zero line on the recording sheet to the preknown depth of the transducer below the water surface, in which case the real distance between the water surface and the river-bed, i.e., the actual water depth is directly recorded. The transportation speed of the recording paper can also

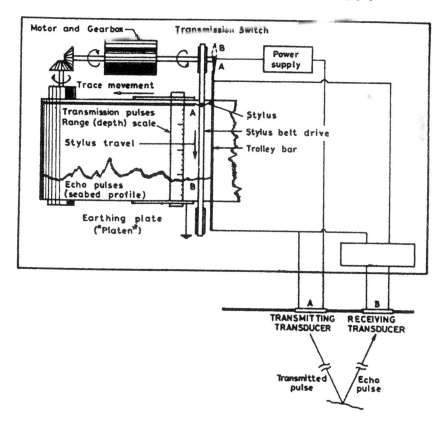

Fig. 6.9   Schematic diagram showing the sounding cycle events

be adjusted. Echo-sounders used for hydrological survey use frequencies of 30, 32, 64, 80 or 210 kHz. With higher frequencies less to no penetration of the sea-bed will take place; only the top of the silt layer or soft unconsolidated bottom material will be recorded. The propagation speed of sound in water is between 1400 to 1550 m/s and varies with density. Again density is influenced by temperature and salinity.

6.9 DATA PROCESSING

The processing of data can be carried out manually or by automated procedure. When using an echo-sounder the following data will be simultaneously recorded on the graph sheet:

  i)   the echo-sounded depth,
  ii) the location of a number of soundings in the cross-section,
  iii) the position of the cross-section or run,
  iv) the time and date.

The position of a number of sounding points is observed at regular intervals and simultaneously plotted on the echo-graph. These plotted marks are numbered. The following survey data is again recorded on standard forms:

  a) Before commencement of survey:
      (i) The data, (ii) the survey locations, (iii) the positioning stations used, (iv) the tidal gauge used.
  b) During the survey:
      (i) The fix numbers of the marked sounding points, (ii) the fix position of the observed positioning system, (iii) the time, regularly at the fix positions.
  c) After the survey:
      (i) The sounded depths as registered on the graph will be reduced to chart datums by introduction of the data obtained from the tide gauges during the sounding operation. The fix positions will be plotted on prepared charts and/or cross-sections. The distance between the plotted fixes on the echo-graph will be divided into regular intervals which correspond with the measured distance intervals between fixes plotted on the chart (Fig. 6.10).

In the case of automatic data-processing the surveyor on board the survey launch will keep a record of the selected position fixes during the sounding run. The marks of the fixes on the echo-graph are numbered on the graphs. The record will state the following information:

  (a) at the start of the sounding run
      (i) the fix number, (ii) the readout of the positioning system, (iii) the time, (iv) the tidal information (b) at the end of the sounding run:

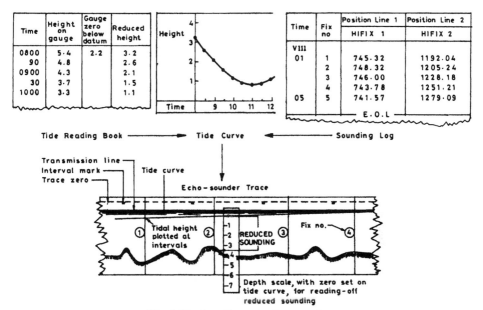

Fig. 6.10 · Sounding reduction procedure

(i) the fix number, (ii) the readout of the positioning system, (iii) the time and (iv) the tidal information.

After completion of the survey the tape with recorded data is removed from the computer and transferred to the survey office for further processing. From the recorded tape the computer in the survey office will produce all the necessary cross-sections, profiles and hydrographic contoured survey charts.

6.10 MAP PROJECTION

Map projection is concerned with the methods of portraying the curved surface of the earth on the plane surface of a map. The salient points to be noted are: the earth is considered a geoid, i.e., a sphere flattened at the poles with an equatorial radius approximately 6378 km and polar radius 6357 km. The position on the earth's surface is designated by geographical co-ordinates, latitude and longitude. As the spheroid surface of the earth cananot be truly projected on a plane, any chart will have some distortion regarding shape, bearing (direction) scale and area.

6.10.1 Mercator Projection

This can be visualised as a projection from the centre of the earth onto an enveloping cylinder tangent at the equator. The Mercator projection is illustrated in Figure 6.11. The axis of the cylinder lies in the plane of the equator.

98

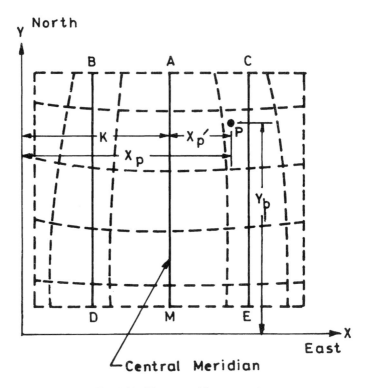

Fig. 6.11 Transverse Mercator projection

The cylinder intersects the spheroid along two circles (BD and CE) equidistant from the central meridian AM. All meridians and parallels are currently as indicated by the broken lines. The central meridian establishes the direction of grid north. The Y-axis is parallel to the central meridian and the X-axis is perpendicular to it.

If the geodetic position of any point P is known, the state-plane Co-ordinates on the transverse Mercator projection can be calculated:

$$Y_p = Y_0 + V\left(\frac{\Delta\lambda'}{100}\right)^2 \pm C \qquad \qquad \dots (6.7)$$

$$X_p = H\Delta\lambda' \pm ab + K \qquad \qquad \dots (6.8)$$

where X_p, Y_p the are state plane co-ordinates of P, $\Delta\lambda'$ difference in longitude between the central meridian and point P in secs, (+)ive when P is east of the central meridian; K the constant distance between the central meridian and the Y-axis.

H, a, Y_0, V are tabulated in United States Coast and Geodetic survey tables vs. Latitude of P, while b, c are tabulated in USGS tables against $\Delta\lambda'$

For lines less than five miles in length the grid co-ordinates may be calculated from the geodetic azimuth with sufficient accuracy for most surveys by the equation:

Grid Azimuth = Geodetic Azimuth $- \Delta\alpha$

where $\Delta\alpha = \Delta\lambda' \sin (\text{Lat. } P) + g$ where $\Delta\lambda'$ is as defined above. Lat P is the latitude of P and g is tabulated in USGS tables against $\Delta\lambda'$. When placing any survey on the state plane co-ordinate system, normal procedure calls for beginning and ending on stations whose state plane co-ordinates are known or can be computed from the known geodetic position. When this is done, all resultant azimuths will be grid azimuths and corrections for convergence of meridian will not be required.

Measured lengths of lines must be reduced to lengths at their respective positions on the state co-ordinate grid. This is most easily accomplished by reducing the measured length to sea level and then reducing the sea level length to grid. The following equation may be used to reduce the measured length to sea level:

$$L_s = L_m \left(\frac{R_e}{R_e + h} \right) \qquad \dots (6.9)$$

where L_s–sea level length; L_m–measured length; R_e–mean radius of the earth (20,906,000 ft); h–average elevation of the line above M.S.L. The sea level length is then multiplied by a scale factor which is obtained from the projection tables for the area on which the line falls.

LITERATURE CITED

1. International Hydrographic Bureau, Supplementary paper-4, special publication, 39, Monaco, Feb, chapter II, pp. 2–4.

2. Accuracy contours, Deice Navigator Company Ltd. System Planning Section, London, Note No: F - 24.

3. Bandyopadhyay, K.K. (1968). An electronic aid to position fixing at Hooghly estuary. National Conf. Port and Harbour Management, Feb. 24-27 1968, Calcutta Port Trust.

CHAPTER 7

PHYSICAL MODELLING
OF TIDAL RIVERS

7.1 INTRODUCTION

Tidal flow simulation in a hydraulic model is governed in principle by the similarity criteria related to the choice of model boundaries and model scales such as in steady flow. Along with that it is necessary to take into account the specific characteristics of oscillatory flow. Before dealing with that a résumé of the modelling laws for rivers in general must be furnished. Hydraulic models are extremely useful in river engineering for understanding morphological processes. The models frunish useful information for suitably imposed hydraulic and physical conditions while designing engineering measures for river regulation works. The model also enables visual examination of the flow pattern. The hydraulic scale model can only be adopted as an aid to design and cannot be considered as 'an end-all' and 'cure-all' to the problems that face the designer.

Nowadays with easily available computer facilities, mathematical modelling has become a very convenient tool for understanding the complex flow situations, prevailing in nature. Mathematical models once calibrated can be a very useful tool for simulation and prediction purposes. Nevertheless, the usefulness of physical models is still very much felt, especially in modelling 3-dimensional flow phenomena. In this chapter scaling laws for distorted and undistorted river models are dealt with so that a hydraulic engineer can design, construct and use physical river models. Particulars regarding water supply system, estimation of pump capacity and arrangement for control of flood and ebb, and water levels at the downstream end, are also included. Rivers are relatively wide, with irregular boundaries and usually exhibits non-uniform unsteady flow. Planning river models means working with a complicated geometry and problem-laden hydraulics. River models with fixed beds are normally run as per Froude's model law, i.e., the ratio of inertial to gravity force concomitantly influences viscosity and roughness. In other words, Reynolds model law, i.e., ratio of inertial reaction to viscous forces needs to be compensated properly. To achieve this model distortion is given, which means giving up exact geometrical similarity in favour of improved similarity of the flow process.

7.2 SCALING LAWS FOR DISTORTED AND UNDISTORTED RIVER MODELS

The scale number of distorted and undistorted river model based on Froude's similarity criterion is shown in table, where n = distortion parameter.

Table 7.1 Scaling laws

	Kinematic or dynamic condition	Derivation of scale	Distorted model $n > 1$	undistorted model $n = 1$
Velocities	$V_r = (L_r/n)^{1/2}$			$V_r = (L_r)^{1/2}$
Times	$t = L/V$;	$t_r = L_r/V_r$	$t_r = (L.n)^{1/2}r$	$t_r = L_r^{1/2}$
Slopes	$I = h/L$	$I_r = 1/n$	$I_r = 1/n$	$I_r = 1$
Reynolds no.	$Re = V\,h/v$	$Re_r = V_r L_r/n$	$Re_r = (L_r/n)^{3/2}$	$Re_r = L_r^{3/2}$
Pressures	$p = F/A$	$p_r = V_r^2$	$p_r = L_r/n$	$p_r = L_r$
Discharges	$Q = V.A$	$Q_r = V_r L_r^2/n$	$Q_r = L_r^{5/2}/n^{3/2}$	$Q_r = L_r^{5/2}$

Model distortion results in higher flow velocities in the model, i.e., increased turbulence, improves similarity as the R_N increases and finally shortens model run time. The scale numbers for fluid properties are $\rho_{wr} = 1$, $n_{wr} = 1$ and so $Re_r = (L_r/n)^{3/2}$.

From the general resistance law for open channel flow, i.e., according to Rouse and Moody [in 6], the scale number for the frictional coefficient can be determined as $\lambda_r = (I_r/F_r^2)$ and with $I_r = 1/n$ and $F_r = 1$, one obtains $\lambda_r = 1/n$.

For derivation of the scale number for relative roughness of the channel the requirement of simulating gravity and frictional forces simultaneously must be met. The empirical formula of Manning-Strickler containing the combined action of gravity, viscosity and roughness is employed:

$$v = 26/k^{1/6}\ h^{2/3}\ I^{1/2}$$

yields $\qquad v_r = k_r^{-1/6}\ (L_r/n)^{2/3}\ (I/n)^{1/2}$ .... (7.1)

Similarity is achieved when

$$v_r = \sqrt{L_r/n} = k_r^{-1/6}\ (L_r/n)^{2/3}\ (I/n)^{1/3}$$

or $\qquad k_r = L_r\ n^{-4}$ .... (7.2)

and with $h_r = L_r/n$, one obtains the scale number of relative roughness

$$(k/h)_r = n^{-3} \qquad \qquad \dots (7.3)$$

Thus the basic concept for simulation of channel roughness in the model is found, i.e., $(k/h)_m = (k/h)_n \, n^3$. In other words, relative roughness of the distorted model must be greater than nature.

7.3 DIMENSIONAL ANALYSIS FOR SEDIMENT TRANSPORT

Coming to the simulation of sediment transport, the specific sediment transport rate g_s expressed in mass per unit time and unit width depends on material properties of fluid and sediment upon grain diameter d of the transported material, depth of flow h and slope I;

$$f\,(g_s,\, \rho_w,\, \rho_s,\, \gamma_w,\, g,\, d,\, h,\, I) = 0 \qquad \qquad \dots (7.4)$$

or by dimensional analysis

$$g_* = f\left(\mathrm{Fr}_*,\, \mathrm{Re}_*\left(\frac{\gamma_w}{\gamma_s - \gamma_w}\right)h/d\right) \qquad \qquad \dots (7.5)$$

where $\qquad g_* = \dfrac{(g_s)}{\rho_s v_* \, d} \quad$ and $\quad v_* = \sqrt{ghI}$.

Again $\qquad \mathrm{Fr}_* = \dfrac{\gamma_w}{\gamma_s - \gamma_w}\, h/d \; I = \dfrac{\tau_0}{(\gamma_s - \gamma_w)d} \qquad \dots (7.6)$

is a modified grain Froude no., i.e., the ratio of sp. wt of fluid and of submerged sediment.

Similarly

$$\mathrm{Re}_* = \sqrt{\frac{ghI\,d}{v_w}} = v_* d/v_w \qquad \qquad \dots (7.7)$$

Dividing $\qquad \dfrac{\mathrm{Fr}}{\mathrm{Re}_*^2} = A_* = \left(\dfrac{\gamma_w}{\gamma_s - \gamma_w}\right) v_w^2/gd^3 \qquad \dots (7.8)$

The dimensionless buoyancy parameter A_* contains exclusively the properties of the fluid and sediment and characterises the material properties of both media. It is free of dependence on either flow or shear stress and so is found to be very useful in the analysis. The beginning of sediment motion exhibits a unique relation between Fr$_*$ and Re$_*$, i.e., the Shield diagram (Fig. 7.1).

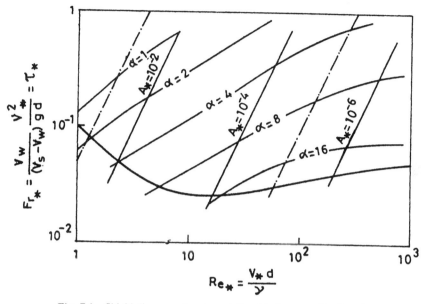

Fig. 7.1 Shield diagram, showing relationship between R_{e*}, F_r and α

7.4 SEDIMENT TRANSPORT EQUATION

The sediment transport relation was developed by Einstein [1], who found that it depends on transport intensity and flow intensity:

$$\phi = \frac{g_s}{\rho_s} \bigg/ \left(\frac{\rho_w}{\rho_s - \rho_w} \right)^{1/2} (1/gd^3)^{1/2} \qquad \cdots (7.9)$$

$$\psi = \text{flow velocity} = \frac{\rho_s - \rho_w}{\rho_w} \, (d/r_{hy}) \, I \qquad \cdots (7.10)$$

For the transport rate ϕ, one obtains

$$\phi = \frac{g_s}{v_* \rho_s d} \left(\frac{\gamma_w}{\gamma_s - \gamma_w} \right)^{1/2} \left(\frac{v_*^2}{gd} \right)^{1/2} = g_* \, \text{Fr}_*^{1/2} \quad \cdots (7.11)$$

and

$$\psi = \left(\frac{\gamma_s - \gamma_w}{\tau_0} \right) d = 1/\text{Fr}_* \qquad \cdots (7.12)$$

So the Einstein bed load function can be represented as a function of g_* and Fr_*.

7.4.1 Motion in Suspension

In turbulent flow the concentration distribution C of the suspended material in the vertical (Fig. 7.2) is described by Rouse [in 1] as:

$$\frac{C}{C_a} = \left(\frac{h-z}{z} \cdot \frac{a}{h-a} \right)^\alpha \qquad \dots (7.13)$$

For C_a at the elevation $(z = a)$ above the bed, Einstein recommends $a = 0.05\ h$ and the exponent

$$\alpha = \frac{1}{\beta} \frac{v_s}{kv_*} = 2.5\ v_s/v_*$$

where β = ratio of turbulent mass to momentum exchange = 1; $k = 0.4$, so α is varies as $\infty\ v_s/v_*$.

The distribution of suspension concentration for different values of α is plotted in Figure 7.3, which shows that at large values of α, i.e., large fall velocity and large shear velocity, suspended material becomes more uniform. With the help of dimensionless buoyancy parameter $A_* = Fr_*/Re_*^2$, a relation between the exponent α of the concentration distribution and parameters Fr_* and Re_* can be formulated. With this, α can be introduced in the Shield

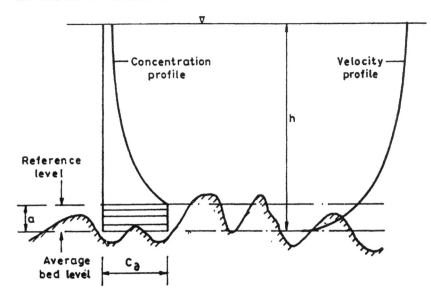

Fig. 7.2 Schematized velocity and concentration profile

diagram. Lines of constant values of α are shown in Figure 7.1. It is thus found that the motion of suspension and concentration distribution in the vertical is determined by the dimensionless parameters Fr_* and Re_*. For large Re_* lines the constant α tends to become horizontal asymptotic, i.e., independent of Re_*. The transport rate g_s for suspended materials in which flow velocities are simulated, is reproduced correctly according to Froude's law if the parameters Re_* and Fr_* for the sediment have equal values in the model and in nature. Accordingly, one obtains:

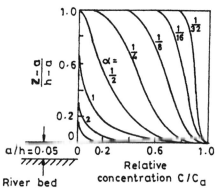

Fig. 7.3 Distribution of suspended load after Rouse

$$h_r^3 = L_r^{3/2} \, \Delta\rho_r \qquad \qquad \dots (7.14)$$

To determine the correspondence between the model scales and the roughness related to grain diameter, the Manning-Strickler equation may be used:

$$(k_{st})r = L_r^{1/2}/h_r^{2/3}$$

For a movable bed the roughness coefficient k can be determined from the grain diameter d_{90} of the cover layer.

Now $\qquad\qquad (k_{st}) = 26/d_{90}^{1/6}$ and with that

$$(k_{st})r = d_r^{-1/6} \qquad\qquad \dots (7.15)$$

where $\qquad\qquad k_{st} = m^{1/3}/s \quad$ and $\quad d_{90}$ in $m.$

Further, one can derive a simple relation between the horizontal and vertical scale $h_r = L_r^{0.7}$, which closely resembles the Lacey-Inglis equation [in 2]. Figure 7.4 shows the derived scale relationship. The chosen horizontal and vertical length scale should intersect on the line for $\Delta\rho_r$ of the chosen sediment. The intersection should coincide with the line $h_r = L_r^{0.7}$.

106

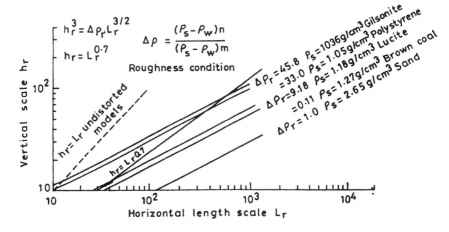

Fig. 7.4   Sediment transport scale relationship

The scale of grain diameter is $d_r = \Delta\bar{\rho}^{1/3}$. The time scale for sediment transport is

$$t_{sr} = (L_r^{3/2}/d_r)\, \eta_r$$

where η_r is the void ratio.

For an identical size curve in nature and the model $\rho_r = 1$. So the time scale for the hydraulic process is

$$t_{sr} = \frac{L_r^{5/2}\, \Delta\rho_r}{h_r^2} \qquad \text{.... (7.16)}$$

Assuming that the flow process takes place according to Froude's model law, then

$$t_{h_r} = L_r/h_r^{1/2} \qquad \text{.... (7.17)}$$

Ergo, the time scale for sediment transport processes and the hydraulic process are not the same in the model.

7.5   CHOICE OF SCALE

Several practical aspects determine the choice of scale for a model of sediment transport. The scales in toto exert an influence on the time scale and hence on the duration of experiments. Further, it should be remembered that the time scale is not a parameter of free choice. The three scale equations can be written as:

$$\text{Re}_{*_r} = 1, \quad L_r^{1/2} \, h_r \, d_r = 1 \qquad \qquad \dots (7.18)$$

$$\text{Fr}_* = 1, \quad L_r^{-1} \, h_r^2 \, d_r^{-1} \, \Delta\rho_r^{-1} = 1 \qquad \qquad \dots (7.19)$$

and roughness condition

$$L_r^3 \, h_r^{-4} \, d_r = 1 \qquad \qquad \dots (7.20)$$

As there are four variables and three equations, one scale can be selected independently and the others determined as shown in Table 7.2. It should be noted that the grain diameter scale is rarely used in practice.

Table 7.2

Chosen scale nos.	L_r	h_r	d_r	$\Delta\rho_r$
L_r		$h_r = L_r^{7/10}$	$d_r = L_r^{-2/10}$	$\Delta\rho_r = L_r^{6/10}$
h_r	$L_r = h_r^{10/7}$		$d_r = h_r^{-2/7}$	$\Delta\rho_r = h_r^{6/7}$
d_r	$L_r = d_r^{-10/2}$	$h_r = d_r^{-7/2}$		$\Delta\rho_r = d_r^{-6/2}$
$\Delta\rho_r$	$L_r = \Delta\rho_r^{10/6}$	$h_r = \Delta\rho_r^{7/5}$	$d = \Delta\rho_r^{-2/6}$	

Free choice of one scale inevitably leads to a distorted model. Distortion is evidenced in an increase in slope, leading to vertical distortion of cross-sections, which in turn results in a velocity distribution that deviates from that observed in nature. Consequent to vertical distortion, one sometimes sees exaggeration of slopes. In other words, the normal slope is increased by a constant slope, i.e., the model becomes tilted. The ratio between length and cross slope is thereby no longer the same in nature and the model. Such a situation is not permissible in modelling of backwater situations and tidal rivers.

7.6 TIDAL FLOW CHARACTERISTICS AND SCALING CRITERIA

The flow characteristics of tidal waters have already been detailed in earlier chapters. Let us only recall here that the smooth sinusoidal variation over the tidal cycle of tide-induced velocities is changed by the coastal zone and geometry of the tidal river/estuary. Figure 7.5 shows the changes that take place due to interaction. Figure 7.6, on the other hand, shows the envelope of high and low water of the simultaneous variation of tidal elevations along the river.

Flow distribution in an estuary over a tidal cycle is generally determined through the cubature method. In the case of tidal flow it is especially important to assume that the flood current velocities slowly decrease during flood tide and that after slack the ebb flow velocities increase in the opposite

108

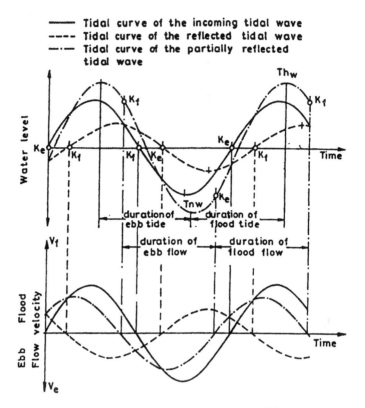

Fig. 7.5 Tidal curves with superposition

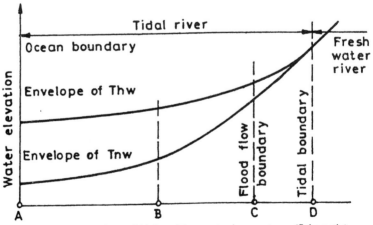

Fig. 7.6 Envelope of high and low water in an estuary (Schematic)

direction (Fig. 7.7). Contrarily, after ebb slack the flood tide velocities increase rapidly and the incoming flood tide reaches high flow velocities very quickly. Scaling is further complicated by surface tension and laminar flow effects for the case of rising and falling water levels in shallow regions, i.e., mud flats. The scale ratio h_r should not exceed 100; the choice of horizontal scale L_r then determines the flow turbulence level, tidal duration, hydrodynamic parameters such as flow velocity and flow rate, and space requirement of the model. A model with a distortion of 5 is capable of producing turbulence levels large enough to allow simulation of prototype conditions during the change in flow direction. Figure 7.8 allows a quick determination of the relations between L_r, h_r, tidal duration and distortion ratio, n.

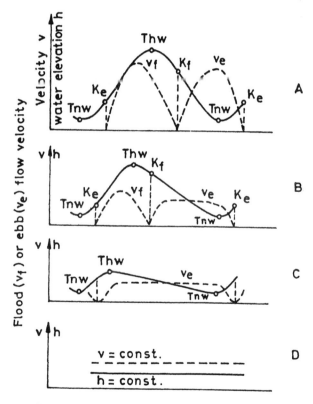

Fig. 7.7 Typical vertical and horizontal tidal curves

For example, with $L_r = 500$, $h_r = 100$ the tidal duration should be approximately 15 min. The basic concepts or model laws for a movable bed were outlined earlier. Here the practical aspects associated with

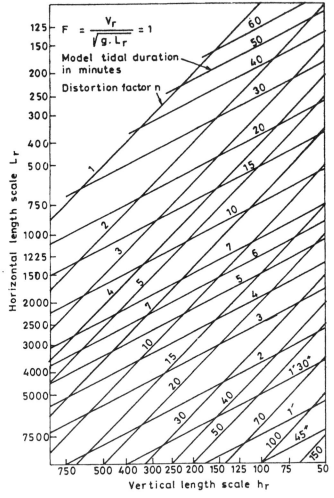

Fig. 7.8   Model scale determination nomogram

determination of model scales for large tidal models can be stated as
follows:

i) L_r and h_r are dependent on space and pump capacities;

ii) h_r also depends on the accuracy of water-level measurements;

iii) diameter and specific gravity of the model sediment depend on the
     material available;

iv) time scale for a prototype duration of 12 h 25 min.

$$L_r = 500, \ h_r = 100 \quad \text{or} \quad n = 5, \ t_r = L_r/h_r$$

$$t_m = \frac{t_n}{L_r/h_r^{1/2}};$$

$$t_m = 745/500 \sqrt{100} = 24.9 \text{ min}$$

If distortion > 5 is used the flow velocities associated with model roughness will be larger than would result from Froude's modelling law. Hence a large duration of a single tide is required to achieve Froude's scaling criteria.

Use of sand as a model sediment is precluded in tidal models because of distortion. Only materials considered to have a density less than sand are used. The following materials have proven satisfactory depending on scaling factors and main hydraulic parameters in the model:

polystyrene (1050 kg/m^3), apricot pits (1329 kg/m^3),

lucite (1185 kg/m^3), bakelite (1500 kg/m^3),

brown coal (1210 kg/m^3) and anthracite (1670 kg/m^3).

Based on the scale grain diameter mentioned earlier, the density and diameter of the sediment material has to satisfy.

$$d_n/d_m = d_r = \Delta\rho_r^{-1/3} \quad \text{and} \quad \sqrt[3]{\frac{\rho_{sm} - \rho_w}{\rho_{sn} - \rho_w}} \qquad \dots\dots (7.21)$$

As explained above, Fr_* and Re_* are the primary parameters for sediment transport. The relations between horizontal and vertical scale can be determined from the material characteristics. The density and diameter of the sediment material must satisfy the interdependence between scale d_r and the specific gravity of the material based on the assumption that $\rho_{sr} = 2600$ kg/m^3, i.e., the prototype material as shown in Figure 7.9. For higher distortion the time factor ξ is to be considered:

$$\xi = \sqrt{h_r/v_r} = \sqrt{\frac{h_n/h_m}{v_n/v_m}} \qquad \dots\dots (7.22)$$

For conversion to prototype condition

$$v_n = v_m \left(\alpha \, h_r^{1/2}\right), \quad Q_n = Q_m \left(\alpha L_r h_r^{3/2}\right)$$

The morphological time scale is a conversion factor to prototype conditions and describes temporal changes of transport phenomena in tidal models, i.e., which time scale should be used in converting the bottom deformations in the model to the prototype. Here one is essentially dependent on experimental results. Based on experimental results a conversion factor of roughly 705 can

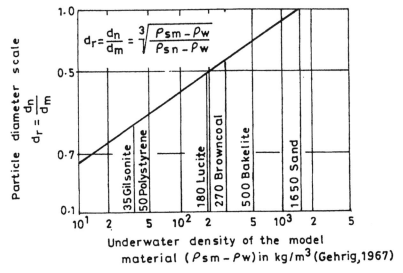

$$d_r = \frac{d_n}{d_m} = \sqrt[3]{\frac{\rho_{sm} - \rho_w}{\rho_{sn} - \rho_w}}$$

Fig. 7.9 Scaling relationship between particle diameter and density of bed material (Gehrig, 1967)

be adopted, which means a model year normally lasts about 1/2 day. The model must be run under given flow conditions for a sufficiently long time to ensure that a steady morphological situation develops. For example, the Elbe model [6] reached a constant morphological condition after 15 yr, i.e., 186 h in the model.

7.7 WATER SUPPLY SYSTEM FOR TIDE GENERATION

The water supply for a tide-generating system is generally provided by continuous pumping of water from a sump with a recirculating system. The pumping capacity is estimated from the maximum discharge requirement of the model. For example, suppose the width of the model at the sea end to be 25 km or 25,000 m, the depth below M.T.L. 10 m and the range of tide 4 m. This means the rise of tide above M.T.L. will equal 2 m. Assume the velocity amplitude as equal to 2.5 m/s. The tidal influx magnitude then equals (width × depth × velocity) 25,000 m × 12 m × 2.5 m/s = 750,000 m³/s. Considering a horizontal scale of 1/800 and a vertical scale of 1/160 (i.e., distortion of 5), the model discharge comes to 0.47 m³/s. So three pumps, each of 0.16 m³/s, may be provided with an identical fourth unit kept ready as a stand-by. To control flood and ebb at the sea end some part of the length has to be reserved for water escape. This calls for the required number of PC-controlled automatic gates of suitable width uniformly distributed and connected to a common drive. Drainage from the model during ebb is effected by free fall over rectangular shaped weirs. The crest level of the weir may be fixed from the weir formula.

$$Q = C_d L H^{3/2} \qquad \qquad \dots (7.23)$$

where, Q is the discharge equal to twice the influx amplitude; C_d coefficient of discharge; L-length; H-height. The numerical value of the height of the water above the weir can be obtained by substituting the relevant values in the discharge equation. The height of the weir has therefore to be kept below the high-water level of the model tray at the sea face by the amount 'height of water' calculated earlier. It may be necessary to adjust the weir height, however, during the proving stage of the model. The capacity of the sump provided must be such as to accommodate water for both tidal influx and initial storage. Tidal influx is normally estimated from the discharge versus time curve. Assuming this curve will approximate a sine curve, the tidal influx will equal

$$\int_0^T A_0 \sin \omega t \, dt \qquad \qquad \dots (7.24)$$

where $A_0 = 0.48$ m^3/s and T = time period per tidal cycle.

The initial storage capacity of the whole model has to be estimated along with that of the inlet and tidal chamber. To this the capacity of the tidal influx must be added. The sum obtained gives the total sump capacity. To ensure clean water, however, by making an allowance for sedimentation, etc., the total sump capacity should be suitably increased. To introduce water at the landward end of the model for the necessary upland discharge, an ingress arrangement is required. For this purpose a suitable connection may be taken from the tidal inlet chamber for the initial filling. Water is distributed from the upstream chamber to the landward end of all the tributaries through the respective labyrinths and individual stilling chambers by means of suitable pipelines. The upland discharge feeds into the system through respective V-notch weirs.

LITERATURE CITED

1. Bogardi, J. (1974). *Sediment Transport in alluvial Streams*. Academe Klado, Budapest.
2. Giesse, E., Harden, H., Vollmers, H.J. (1973). The tidal regime of the Elbe-River. Hydraulic Model with moveable bed. *Proc. Int. Symp. River Mechanics*. Bangkok, Thailand.
3. Giesse, E., Harden, H., Vollmers, H.J. (1974). Experience with moveable bed tidal models. Proc. *14th Int. Conf. Coastal Engineering*. Copenhagen, Denmark.
4. Giesse, E. and Vollmers, H.J. (1975). On the reproduction of morphological changes in a coastal model with moveable bed. *Proc. XV Int. IAHR Cong*. Sao-Paulo, Brazil.
5. Giesse, E. (1976). Stability problems for the navigational channel in a tidal river. *Proc. River 76*. Fort Collins, Colorado.
6. Kobus, H. *Hydraulic Modelling*. German Assoc. *Land Resources and Land Improvement, Bull*. 7. Pitman Books Ltd., London.

CHAPTER 8

HYDRAULIC ASPECTS OF RIVER TRAINING WORKS IN TIDAL RIVERS

8.1 GENERAL

The training of a river involves construction of structures across or along a stream to achieve certain objectives. These include levees built along the length of the stream to contain floods, and spurs and guide-banks to alter the local flow conditions, in other words to guide the flow. Dredging is likewise extensively resorted to in training a river. mainly for navigational purposes, but also to divert the flow into secondary channels or to execute cut-offs of the main stream to reduce the flood level. The various alterations and river training works further include measures for protection of banks, either directly through pitching or indirectly by construction of spurs.

The design of bank protection works and river training depends on the area of the river as well as the purpose for which they are undertaken. The objectives of river training are flood control and protection, navigation, sediment control and guidance of flow. A river can usually be divided into four distinct reaches: (i) the reach where the flow takes place under the action of gravity in one direction only, also known as the non-tidal reach; (ii) the tidal component from the upper tidal limit to the location where the effect of fresh upland discharge ceases. This reach has well-defined banks within which the flood and ebb tides flow; (iii) the estuary portion where the shorelines open out. Here the channels are maintained by tidal currents; (iv) the estuary mouth where maintenance is governed by the combined effects of waves and tides.

Rivers normally carry a huge quantity of run-off waters from the catchment area along with suspended and bed loads in the non-tidal reach. If the river is not fully charged with sediment, it picks up material from the bed and erodes the banks depending on local conditions. The material is then dropped further downstream in slack-water regions. This process of deposition goes on in one direction only, however, which may develop undesirable curvatures which can be set right with suitable training measures.

8.2 CHARACTERISTICS OF TIDAL RIVERS AND ESTUARIES

In the upper tidal reaches of a river the silt-laden water enters the tidal zone. This creates a complicated scenario since the tidal current and the downward freshet are opposed in direction. With high river discharge, the energy of the opposing tidal current may be less and thus yield a resultant current in the downward direction. If the process continues in a sufficiently long tidal reach where the downward currents meet with sufficiently strong tidal currents, the effect is to push down the tidal limit. Thus the tidal limit would be further downstream during neap tides compared to spring tides.

In India the upland discharge decreases during the nonmonsoon or dry season. Therefore, ebb currents begin to predominate. As the same volume of water entering the tidal river during the flood has to go out, it is apparent that the greater the influx, the greater will be the scouring power of the currents. It is thus most important in tidal river training to ensure maximum flux. As one proceeds downstream the rate of influx gradually increases and consequently the effect of upland water decreases until a point is reached below which the influence of the upland discharge disappears. So the channels below this point are not governed by ebb; they are predominantly flood maintained.

In the wider portion of the estuary the path followed by flood currents differs from that followed by ebb currents. Local changes accompanied by changes in flood and ebb currents exert a greater effect in maintaining the equilibrium of the system. In particular, channels carrying the main ebb currents at low tide are generally dominated by the ebb with net movement of sediment on the bed directed seaward. They are terminated by a shallow bar at the seaward end of the shoals that form outside the estuary.

Flood-dominated channels, on the other hand, are usually terminated by a shoal at the landward end. They can be readily identified from the chart of the estuary. Considerable sediment is in circulation between the two channels; any realignment affects this circulation and may lead to deposition of sediment in those parts of the system that were previously in dynamic equilibrium. Such effects must be taken into consideration while planning river training measures.

Reduction of intertidal volume anywhere in the estuary reduces the tidal influx through every cross-section seaward of that point. If the reduction is small, it exerts a rather insignificant effect on the water levels. So the reduction in flow is equal at all cross-sections but the effect is great immediately downstream of the reduced part. In practice, a large reclamation affects the tidal range so that changes in tidal influx vary from section to section. In any estuary where the alluvial channel is in dynamic equilibrium, reduction of intertidal volume has dangerous consequences. The weakening of currents seaward of the reclamation work causes deposition of sediment. The effect is greatest when the reduction is made near the upstream tidal limit, resulting in serious siltation. Movement of a tidal wave in shallow water results in retardation of the low-water part of the wave relative to the high-water part. The wavefront steepens progressively as it advances, which results in the time of rise becoming shorter than the time of fall of tide. The resultant increase in flood velocities over ebb causes a strong landward bias in the direction of sediment transport. Tidal propagation can be made more rapid at all states of the tide provided the hydraulic mean depth of the channel can be deepened.

8.3 IMPROVEMENT OF TIDAL RIVERS

Certain principles are to be observed while designing improvement works in tidal rivers. Physical characteristics such as tide at the mouth, silt content, shore irregularities, cross-sectional area, average width and depth, bed features, upland discharge, etc., need to be studied carefully before deciding the type of training measures to be implemented. The principles to be observed are:

i)   the river and estuary should be considered in their entirety:
ii) improvement works should obtain a general balance of all the forces in operation and be designed insofar as possible in accordance with natural tendencies;
iii) divergence between flood and ebb axes should be minimised as far as practible to obtain maximum benefit from the flood and ebb energies;
iv) entrance of the tidal wave from the sea should remain unhindered and waterways uniformly diminished in width from the mouth onwards;
v)   velocities and tidal volume, which increase with tidal range, are the most important agents in maintaining depth of the navigational channel;
vi) improvements should commence from the lower reaches of the river upwards.

8.4   TRAINING MEASURES

It is necessary first of all to compute the cross-sectional areas, widths and mean depths at midtide from the survey charts. Average curves may then be drawn for the cross-sectional areas and depths to work out empirical relations, if any. A convenient way of ascertaining relative changes is to make time-hsitory diagrams of the cross-sections. The method involves the following steps.

Take a time line along the vertical axis in a suitable scale. Then draw parellel lines for the cross-sections to be compared. The depths that have occurred at different times are written down along these horizontal lines. contour lines are now drawn at suitable depth intervals (see Fig. 8.1).

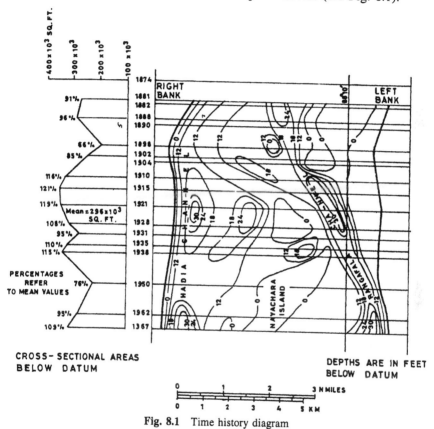

Fig. 8.1 Time history diagram

Should the axes of the flood and ebb be divergent (Fig. 8.2), a careful study of the strength and direction of currents in the river reach is required, since sufficient depths are not available when the energy of the flood and ebb currents is not concentrated in a single channel. The main aims of tidal river training are to obtain a gradually reduced section upstream, to contract the river by longitudinal training walls and to make flood and ebb axes coincident by constructions of suitable groynes, closing of channels, etc. A tidal river splits into a number of channels in the estuary region. Since training seeks to create a channel maintained by flood currents, the orientation and alignment of these channels are of paramount importance. Tidal currents alone can maintain a channel if it is sufficiently deep and favourable to flood and ebb flow.

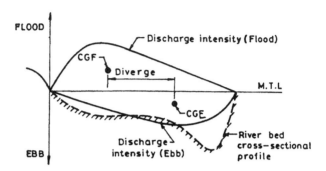

Fig. 8.2   Divergence of flood and ebb flow

8.4.1   Training Measures for River Mouth

To improve river mouths, one has to take into consideration another factor, namely wave action, which plays a significant role in the formation of bars in the reach, especially during heavy storms. Bars are also formed due to alluvial material brought down from the river when the tide is weak and upland discharge is high. In this case the conditions at the outfall are controlled by littoral current and discharge, which results in the formation of the river delta. Depending on the alluvial material brought down by the river, bars may be hard or loose. Hard bars can be removed by dredging and blasting.

Improvement of river mouths requires installation of artificial works such as training walls or jetties to increase velocities for carrying the river load farther down. The load will either go directly into the deep sea or be moved by the littoral or other current system. Conversely, a high river discharge emptying into the sea may be arrested on its way by a littoral current. This will result in the formation of a sandspit parallel to the coastline, through which the river can break during flood. Control of littoral drift is the main problem in channel improvement.

This problem can be resolved by installing a jetty. The jetty would cause shoaling in the updrift side (Fig. 8.3), which would then begin to pass around the other end of the jetty to be picked up by tidal currents. Where change of drift direction is expected, it is desirable to construct a jetty on both sides of the harbour.

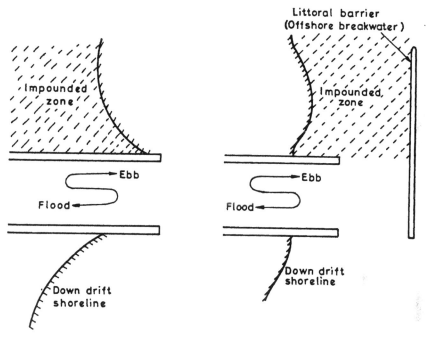

Fig. 8.3 (i) Jettied inlet and (ii) jettied inlet with offshore breakwater

8.5 CASE STUDY

Figure 8.4 shows the Sankrail reach of the Hooghly River. The river is very wide in the middle and narrow in the upstream and downstream directions. during high monsoon discharge the strong ebb would scour the bight on the right bank where adequate navigational depths could readily be obtained. In the dry season the flood current would not follow the bight and tended to develop a cul-de-sac on the sand. This resulted in deterioration of the navigational channel.

To resolve this problem, a groyne was constructed at Akra based on CWPRS recommendations; it effectively minimised the divergence between the flood and ebb axes by repelling the flood flow.

120

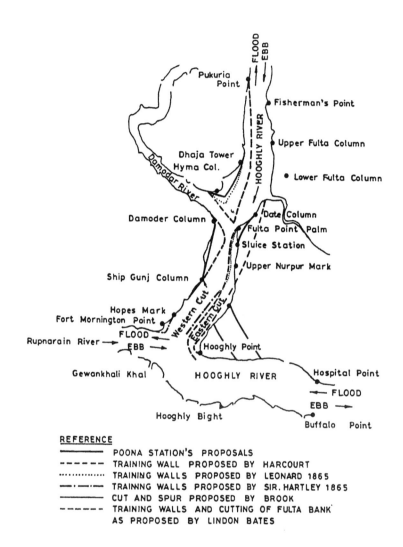

Fig. 8.4 Plan of the Hooghly River showing proposals for improvement of the Fulta-James and Mary Reach by different consultants

LITERATURE CITED

1. River Behaviour Management and Control. Central Board of Irrigation and Power, New Delhi, publ. no. 60 (1989).
2. Roy, S.C. (1969). Hydraulic Investigation on Behalf of Hooghly Estuary. Sonderdruk aus Heft 32 der Mitteiulungen, des Franzius Institute fur Technischen Universitat, Hannover.

CHAPTER 9

BANK INSTABILITY AND EROSION CONTROL MEASURES

9.1 INTRODUCTION

Caving of bank is one of the principal reason of deterioration of river conditions. Bank protection therefore forms an integral part in the river training programme. Rivers passing through heavily populated areas require protection from erosion of valuable lands and properties. Similarly, protection against erosion means often saving the holdings from destruction. Banks located upstream and downstream of hydraualic structures such as weirs, barrages and bridges have to be protected depending on the direction and nature of current. Further, bank protection is needed for flood embankments which on breaching will cause disaster to the large areas protected by them.

9.2 OBJECTIVES OF BANK PROTECTION MEASURES

To summarise the purposes for which bank protection is required may be any one or all of the following:
  (i)   Protection of hydraulic structures,
 (ii)   Protection of lands where valuable properties are located as well as agricultural lands,
(iii)   Protection of flood embankments and for navigational requirements, and finally by river training.

9.3 CAUSES OF BANK INSTABILITY

Generally speaking, the instability of river bank is the cause of bank erosion. River bank may be damaged basically in two ways: (a) direct removal of material at the surface by scour; (b) internal shear failures resulting in sudden caving in or sloughing of large bodies of earth. Such shear failures may be caused by: (i) surface erosion at the toe of the slope, (ii) general bed deposition, (iii) excessive saturation of bank at low water, (iv) slope angle being too steep and earthquake effect.

When the water level rise with increase in the flood stage, banks become partially or fully saturated with water. For silt or silty sand the angle of shearing resistance may be as low as 50 percent of its original value before saturation. If the angle of the sloping surface is steeper than the reduced angle of the shearing resistance sloughing may result. Seepage is also dangerous during high flood stage if the bank is in the form of an embankment and the landward slope is too steep for the seepage to be contained in the cross-section.

On a meandering river the outside bank is eroded by local high velocity near the bank and at the same time deposition on bars occur in the inside bank. The factors responsible for erosion can be classified broadly as due to (i) waves, (ii) current, (iii) discharges of ground and surface water, water level fluctuations etc. Waves are formed due to wind action, movement of vessels and as well as formation of tidal bores. The current causes frictional drag on bed and bank materials. Run-off from surface causes bank erosion where proper surface drainage system is not there.

9.3.1 Identification of Cause of Failure

Proper identification of damaged banks whether due to scour or shear failure is essential as the treatment is different. If scour is the problem the protective measures could be groynes, revetment or vegetation cover so as to keep flow with scouring velocity safely away from the bank materials. If sliding is the problem the embankment slope should be reduced or an intermediate berm be provided so as to increase the stability of soil. Alternately, the soil can be compacted to improve shear strength or drainage condition be improved to reduce seepage pressure following rapid draw down. Normally flow condition met in the river system is turbulent over rough boundaries for which established velolcity distribution law exists. Apart from that what is important to know is the behaviour of velocity structure in the vulnerable zone before undertaking measures for protection.

The river bank consists of the upper and lower sections. The lower bank, the part below low water act as the foundation for the support of the upper bank and is more susceptible to erosion. Recession of the bank is caused by the erosion of the lower bank particularly at the toe. The recession will be fast specially when there is a sandy sub-stratum below, sand is normallly washed away by a strong current and the overhanging bank collapses. The upper bank is the portion between the low water and the high water. Here the erosion is severe where the current impinges normal to the bank. During high stages of the river erosion is, also caused due to a strong current along the bank.

9.4 BASIC CONCEPTS OF EROSION CONTROL

The basic premise of erosion control measures rests on maintaining an acceptable profile of the water course with permissible tolerances. The control measures therefore aim at (i) inducing accretion and (ii) arresting migration of bank or bed materials as the case may be. Planning can only be done after due assessment of the following:

(i)   Extent of onshore-offshore sediment transport and its seasonal variation

(ii) Direction of long shore drift, its seasonal variation and the ratio of net to gross movement

(iii) Factors such as water level variation, change in sediment supply, annual volumetric loss of sediment, and

(iv) Analysis of differentials and extreme event statistics. A designer after analysis of all the prototype data is supposed to specify first the minimum bank gradient necessary to ensure the required level of protection through a seasonal cycle before recommending specific control measures.

9.5   PREPARATION OF THE SCHEME

For the preparation of project report on bank erosion the necessary minimum investigation required are as follows:

(a) Cross-sections of the river where the bank protection works are envisaged for at least 2 to 5 years and the cross-sections should cover all the flow events.

(b) Topographical survey showing the alignment of the river banks and flow path of the river extending at least 1000 m each both upstream and downstream.

(c) Gauge and discharge observations at the proposed site for collection of gauge and discharge data at or nearby site for a period of at least 5 years.

(d) Preliminary soil investigation for type of soil, its gradation, permeability, shear strength etc.

(e) Collection of bed samples and bank materials for analysis of silt factor.

The various methods of river bank protection works are: (i) slope protection and bank along with toe protection against scour, (ii) Flexible type of screen, (iii) construction of porcupines, (iv) construction of spurs. The other erosion control measures are: (a) off-shore breakwater; and (b) artificial headlongs.

9.6   BANK PROTECTION BY RIPRAP

The variables related to bank erosion which affect geometry and stability of channels bed forms in sandy and gravel bed channels such as velocity of flow are interdependent. Some of the variables change with the conditions of flow and alter their roles from dependent to independent variables. In field studies specially it is difficult to distinguish between dependent and independent variables.

The principal variables involved in the analysis of flow in alluvial river channels are:

$$V = \text{fnc } (d, s_e, \rho, \rho_s g, D, \sigma_s, s_\gamma, s_c, f_s, w) \quad \dots (9.1)$$

where V = average flow velocity, d = average depth, s_e = slope of energy grade line, ρ, ρ_s density of water and sediment particles, D = representative fall diameter of the bed material σ_s = measure of the size distribution = D_{50}/D_{15}, s_r, s_c = shape factor of the reach and cross-section f_s = seepage force in the bed of the stream, w = fall velocity of the bed material, g = acceleration due to gravity

9.6.1 Characteristics of Bank Types

Different types of banks have different erosion characteristics. Non-cohesive banks are subjected to surface erosion which results piecemeal loss of material over a period of time. The surface erosion is affected by: (i) direction and magnitude of flow velocity adjacent to the bank; (ii) fluctuations of the flow due to turbulence; (iii) magnitude and fluctuations of shear on the bank, seepage, piping and wave forces.

In the case of soils made up of fine and well-rounded particles the angle of repose is about 20 degrees i.e. (1 V: 2.5 H). For large angular particles it is about 45 degrees.

Fine-grained cohesionless soils drain more slowly than coarse-grained cohesionless soils and are therefore more susceptible to failure. Loose sandy material containing some fines is subject to liquefaction failure i.e. flow slides. When such banks are saturated and the water level drops pressure in the bank builds up to where the sand particles slip. The soil mass then act as a liquid with large quantities of bank material flowing into the channel.

Cohesive banks has a tendency to drain slow because of low permeability. They become unstable under conditions of rapidly falling stage. Accordingly, they are subject to sudden failure by bank caving, sloughing or sliding. When the banks are under cut or saturated large amount of material can be eroded into the channel almost instantaneously. Bank height is important in assessing stability of cohesive banks because such banks tend to be more unstable than non-cohesive banks or banks of low cohesivity of equal height.

In the case of stratified banks the erosion process is complex. The layers of non-cohesive material are subject to surface erosion at the same time they are partially protected by cohesive material. They are therefore subject to erosion and sliding as a result of subsurface flow and piping. Subsurface flow from the bank to the river can occur due to rise in groundwater table, fluctuation in river stage and waves due to wind or ships. When subsurface flow through the permeable layer transports and removes particles from the permeable layers, piping occurs, which causes the overhanging layers to drop and crack. It then becomes vulnerable to erosion from surface flow sliding.

9.6.2 Design for Protection by Stone Riprap

To protect the eroding bank pitching with stone riprap and for protection of toe launching apron is resorted to. The important factors to be considered in the design for protection by stone riprap are:

(i) Durability against abrasion and breaking due to impact by debris brought down by flowing river,

(ii) Size, density shape and angularity of stones,

(iii) Angle of repose of stone and the slope of the bank line,

(iv) Velocity of flow in the vicinity of stones,

For the determination of size and weight of stones various formulae can be used:

Army Corps of Engineers (USA)

$$W_s = \frac{C_D^6 \, \gamma_s \, V_K^6 \, \gamma_f}{8g^3 \, \gamma_s^3} \qquad \dots (9.2)$$

where W_s = weight of bed riprap element, γ_s = unit weight of stones, C_D = drag coefficient whose value is taken as 0.45 for cubic stones. V_K = velocity acting on the stones, γ_s submerged unit weight of stones, g = acceleration due to gravity. The formula is in FPS units and is meant for river bank protection.

Isbash [1] proposed the formula for construction of dams by depositing rocks in running water. The formula takes into account the slope of the bank.

The formula is

$$W = \frac{0.02323S_g \, V^6}{(S_g - 1)\cos\phi} \qquad \dots (9.3)$$

here W = wt. of stone in pounds,

S_g = specific gravity of stones,

V = velocity acting on the stones,

ϕ = angle of the pavement with the horizontal.

The IS-8408, 1976 [2] indicates methods for determination of size of stone for a given velocity. The figure provides two separate curves one for isolated stones and the other for the surrounded ones.

$$W = \frac{(0.02323S_g \, V^6) \, K}{(S_g - 1)^3} \qquad \dots (9.4)$$

$$K = \left(\frac{1}{1 - \dfrac{\sin^2\theta}{\sin^2\phi}} \right)^{1/2} \qquad \dots (9.5)$$

126

where, W = wt of stone in kg; S_s = specific gravity of stones, ϕ = angle of repose; θ = angle of sloping bank, V = velocity in m/s.

The important parameters to be considered while deciding the value of weight stone for bank protection, are respectively specific gravity, bank slope, gradation of stones and its curve, thickness of pitching.

The specific gravity of stones has a bearing on selection of stone weight. The formulae assumes a value of sp. gr as 2.65. However, in actual practice the specific gravity of stone can vary from 2.0 to 3.0 depending on the availability of local material. The effect of specific gravity for determination of velocity as a function of velocity is:

$$W \propto \frac{S_s}{(S_s - 1)^3} \qquad \qquad \dots (9.6)$$

The variation in the weight for the same velocity against the weight of stone required for protection under different specific gravity of the material is shown in Fig. 9.1.

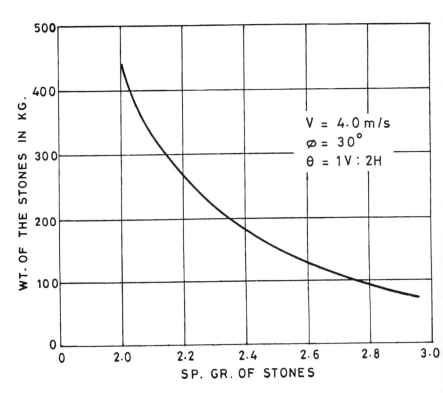

Fig. 9.1 Variation in wt. for the same velocity against wt. of stone under different specific gravity of the material

The stability of the stone on inclined plane is reduced due to its self-weight component which helps to destabilize and tends to roll on the inclined plane. Thus it would be seen that even under condition of no flow, the dumped rip rap of stone will be stable at its angle of repose. Therefore a general formula should take care of the bank slope and angle of repose of the stone used for bank protection. The ISI standard and IRC i.e. Indian Road Congress code assumes a slope of 2;1 for loose boulders and 1.5:1 for concrete block pitching. Effect of variation in stone weight required for various bank slopes is shown in Fig. 9.2. It shows the graph of stone weight and bank slope. The asymmetric shape of the curve approaching towards the angle of repose may be observed.

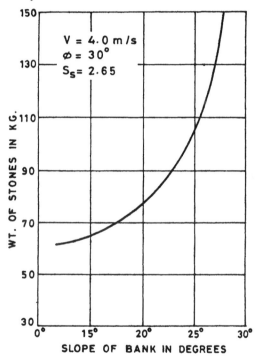

Fig. 9.2 Variation of stone weight for various bank slope

The gradation of riprap should follow smooth size distribution curve. The ratio of D_{max} to D_{50} and the ratio of D_{50} to D_{20} should be around 2. Such a distribution of size should help to fill up the interstices formed by large stones and better interlocking. That will also ensure better resistance to erosion action.

The pitching thickness should be equal to the largest size of stone for well graded riprap. This should be about 1.30 times the average size of

128

stones. Under stronger current 50 percent increase should be considered for the design. The thickness may be checked up with the formula proposed by Spring [3].

$$T = 0.06Q^{1/3} \qquad \qquad \dots (9.7)$$

where T = thickness of pitching in m; Q = discharge in m³/s.

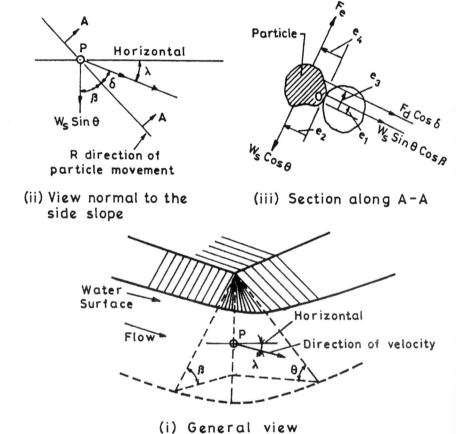

(ii) View normal to the side slope

(iii) Section along A–A

(i) General view

Fig. 9.3 Stability condition for riprap

To determine the factor of safety of stone riprap on slope it is assumed that the riprap is standing under water depth with an angle θ made with the horizontal, ϕ is the angle of repose of the material used for riprap. D is the diameter of stones, β = angle of the vertical made with the direction of resultant particle movement and δ is the angle made by the direction of flow with the direction of particle movement, λ is the angle made by the velocity vector with the horizontal, S_s is the rock specific gravity. With reference to Fig. 9.3 it can be shown that the stability factor $S.F.$

$$S.F. = \frac{\cos \theta \tan \phi}{\eta^1 \tan \phi + \sin \theta \cos \beta} \qquad \dots (9.8)$$

where

$$\eta^1 = \eta \; \frac{1 + \sin (\gamma + \beta)}{2}$$

$$\eta = \frac{2 I \tau_0}{(S_s - 1) D \gamma}$$

$$\beta = \tan^{-1} \left(\frac{\cos \gamma}{\dfrac{2 \sin \theta}{\tan \phi} + \sin \gamma} \right)$$

9.7 SCOUR PROTECTION AT TOE

It is necessary for bank riprap protection to provide stone protruding at the river bed against scour. Due to scour at the toe there could be undermining and collapse of the stone pitching. For this purpose a stone cover known as apron is laid beyond the toe on the horizontal river bed so that scour undermines the apron first starting at its farthest end and works backwards towards the slope. The apron then launches to cover the face of the scour with stone forming a continuous carpet below the permanent slopes of the bank. Adequate quantity of stones for the apron has to be provided to ensure complete protection of the whole of the apron thickness, depth of scour and the slope of the launched apron. For detail design purpose reference may be made to publication of river behaviour control and training published by CBI&P [3]. Figure 9.4 shows typical apron protection measures. The stone size for launching apron may also be obtained from the formulae mentioned. It should be considered having a slope of 2:1 and the thickness of the apron after launching 1.25T. To prevent loss of materials it is advisable to put the launching apron at the LWL = 1.5 D, where D = depth of scour level from the LWL.

Spring also recommends that this laid apron should have a wedge shape with a thickness of T at the inner end with the junction of the bank pitching and increased upto 2.76 T at the other end, T being the thickness of pitching in m. For sudden deep scour spring recommends 1.5 to 2.25 T.

The scour depth calcualtion is based on Lacey's [4] relationships. They are for looseness factor >1.0

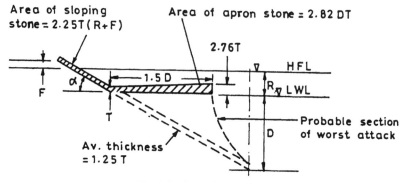

Fig. 9.4 According to spring

$$R = 0.473 \left(\frac{Q}{f}\right)^{1/3} \qquad \qquad \dots (9.9)$$

and for looseness factor < 1.0,

$$R = 1.35 \left(\frac{q^2}{f}\right)^{1/3} \qquad \qquad \dots (9.10)$$

where,

R = scour depth below H.F.L. in m, Q = high flood discharge in m³/s;

f = silt factor = 1.76 $\sqrt{d_m}$, d_m = particle mean diameter in mm;

q = intensity of discharge in m²/s.

The looseness factor is the ratio of width provided to the theoretically computed minimum value obtained by using Lacey's regime equation. The value of scour depth to be used in design are, straight reach 1.25 R moderate bend 1.5 R and for severe bend 2.15 R.

9.8   CONSTRUCTION OF SPURS OR SPUR SYSTEM

Spur or spur systems alter bank line orientations by deflecting currents. Orientation of spurs to the bank may be different considering the flow conditions, bathymetric situations and the objective. Spurs may be of different shapes i.e. straight, *Y-L* or *T* shaped. Length of spurs is governed by the range of water level fluctuations and bank shape. Height is determined by the local scour level, is dependent on the soil compaction and sizes of bank material. Height of a spur may need on course adjustments with change in flow conditions. Spacing between the spurs depends on the steepness of bank slopes, steeper the bank closer should be the interspacing. Consideration should be given to the estimated variation of bank slope in between the spurs. End spurs or terminal spurs are usually higher than the rest in order to promote larger accretion for nourishment of depleted banks. In the case of alluvial soil, range of spacing to length ratio of spurs varies from 0.8 to 2.7

9.9 CONSTRUCTION OF BANK PROTECTION MEASURES

The construction of bank protection measures is done over a two stage lining. First the construction of strong and robust outer protection layer to absorb and dissipate the wave energy and resist the dislodging effects of currents. The second stage comprises construction of an inner protective core or layer. This layer is essentially a filter layer which has to perform two main functions i.e. penetration of high differential water pressure across it and arresting undue loss of bank materials. For the bed protection the filter layer on the river bed is suitably ballasted to keep the composite construction in position against two major loads i.e. hydraulic and mechanical. The hydraulic load include gradients, excess differential water pressure, permeability, rates of filtration and conditions of flow apart from current velocities and waves. Mechanical loads include load on top, grain and water stresses, compaction settlement and shear resistance. It will be dominant when the soil is undergoing deformations for various reasons.

9.9.1 Use of Synthetic Filters

Nowadays synthetic filter or geotextiles are gradually replacing other types of filters mainly because of the fact that they can exactly fulfil the functional requirments of filter. Production of geotextiles can be suited to comply with the various mechanical requirements. There are two main types of geotextiles woven and non-woven type. A woven fabric is a flat structure of at least two sets of threads, one set runs in a lengthwise direction (called wrap) and the other across it called as wept. There are various weaving patterns giving rise to a number of different fabric structures suited to various requirements.

A non-woven geotextile is a textile structure produced by bringing together the fibrous material through mechanical means (needling) or thermal process (partial melting) or chemical methods (gluing) or through a combination of any of these methods. Two basic functions are to be performed by geotextiles for erosion control. They are: (i) retention of base material, (ii) prevention of differential over pressure. The retention of base material is sometimes known as geometrical sand tightness and is independent of the magnitude of hydrostatic load. The flow conditions to which geotextiles is subjected to steady and cyclic flow to the plane and perpendicular to the geotextiles. Rational design criteria are under study based on the premise that there is similarity between flows in open channel and through the pores of a filter at the threshold of sediment transport.

All types of geotextiles allow water flow in the enplane and transverse directions in a widely varying degrees. The permeability requirments are usually based on the hydraulic headloss through the geotextiles and the consideration that the increase in headloss would not affect the design performance. Clogging of soil particles into the pores of the geotextiles often impairs its permeability and should be taken into account while fixing the permeability criteria of geotextiles. Geotextiles used for erosion control should be able to withstand puncture and tearing. Punching is caused from the impact of the falling boulders on the geotextiles and is a function of effective boulder weight, height of placement of boulders, shape of boulders particularly the sharpness of edges, slope-based soil conditions and nature of protective cushion provided over the geotextiles. Geotextiles to be used for bed protection would require much higher tensile strength. The following criteria for woven geotextiles may be adopted:

Warp-4.0 KN/sq. cm and weft 2.5 KN/sq. cm. The other factors to be considered before selecting the type of geotextiles are blocking and clogging. By blocking is meant partial closure of the pores of the geotextiles by grains from adjacent soilmass and thus reducing the water permeability of the geotextile.

Clogging means deposition of the silt in the fabric meshes due to growth of algae, foreign precipitation, etc. causing decrease in permeability.

9.10 CAST STUDY I: BANK PROTECTION BY USE OF GEOTEXTILES AND RIPRAP

At the estuary of river Dadhar near Gandhar in Baruch district of Gujarat, the river embankment was eroding fast due to action of tidal water specially during new and full moon days. The breaches were observed at many places which are enlarging the active oil wells of the ONGC, which are situated on the bank. The erosion rate was estimated to be about 10 m per month. The soil characteristics in the area was found to be a very fine, silty and non-cohesive clay. In this case use of nonwoven geotextile as filter media to check the movement of fine clay/sand was considered to be appropriate. In order to keep the geotextile in place gabion filled with rip-rap were used near the entire fabric area. Figure 9.5 shows the protection works where the non-woven geotextile was laid over a prepared slope having a gradient of 1 : 1.5. The total area covered by geotextile is 2250 sq.m. The fabric was directly spread over the wet subgrade along the entire embankment from toe edge over the slope and the adjacent fabrics were overlapped by 3.0 cm at every 3 m distance. After properly overhanging the fabric at the top edge and at the toe of the embankment about 5 cm thick of river sand layer was spread over the top of the fabric. The gabion made of high density polyethylene type prefabricated into a box-like structure with dimension of 2 × 1 × 1 in m were placed at the toe of the embankment side by side and latter filled with riprap. Taking the support of the toe gabion, other gabion were placed on the slope which were also filled up with riprap.

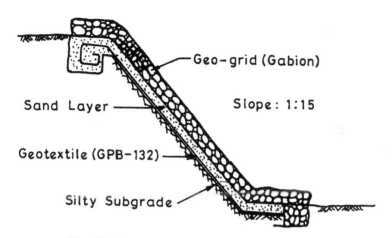

Fig. 9.5 Bank protection using riprap and geotextile

9.11 CASE STUDY II: BANK PROTECTION BY CONSTRUCTION OF PORCUPINES [5]

Porcupines have been found to be effective in case of tidal rivers where silt and clay are available. The porcupines also dampen the river flow and allow deposition of silts which ultimately strengthens the bank.

Tidal channel Natsal on the Rupnarayan river is an example of unstable channel which is evident from the superimposed cross-section of the river for the years 1955, 1965, 1968, 1972 and 1985 (Fig. 9.6). There is a char towards the right bank within the cross-section which is submerged during peak flood and also divide into two distinct streams Fig. 9.7, the right bank channel is deep and depth = 8 m, while the channel on the left side is wide width is about 750 m. From the figure it can be seen that the left bank portion has remained unchanged while the right bank had shifted outside from year to year encroaching adjacent valleys. The tidal elevations and mean velocity for the locations is shown in Fig. 9.8.

134

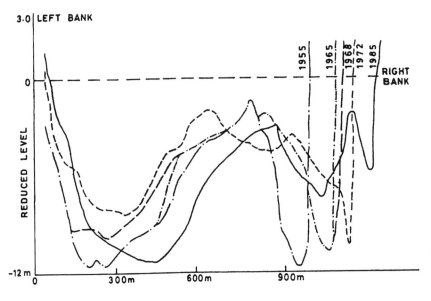

Fig. 9.6 Superimposed cross-section

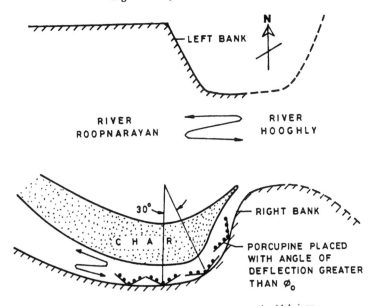

Fig. 9.7 Bank protection measures in tidal river

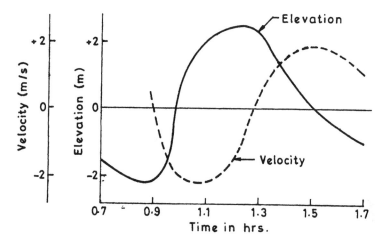

Fig. 9.8 Tidal curve at the location

In the deep and narrow meandering channel spiral flow takes place and no precaution against it is required. For the protection against scour one has to examine whether the critical shear stress, τ_0 or bottom velocity is sufficient to erode the material.

The velocity distribution law can be expressed in the form:

$$u = \frac{u_*}{k} \log_e \frac{y}{y_0} \qquad \qquad \text{.... (9.11)}$$

where, u = velocity at distance y above the bottom, y_0 = dynamic roughness length which represents the distance above the bottom where the velocity is zero, k = Von Karman's universal constant = 0.4, u_* = shear velocity which is constant in the boundary layer, = $\sqrt{\tau_0/\rho}$, where τ_0 is the bed shear stress and ρ density of water,

u and y can be determined from simultaneous current speed measurements at several heights above the boundary by regression of ln y in u which requires several simultaneous measurement in the bottom of the water column. In the case of comparative straight channel having moderate meander where the bottom velocity or bed shear is not large due to flatter bed surface slope, bank erosion may still take place due to wave action mainly during monsoon such as in the case of Sunderbans estuary bank. Proper protection is required and for it use of porcupines for damping out spiral flow as well as for initiating siltation at the concave bank and assisting the same at the vulnerable zones is done. Porcupines Fig. 9.9 with bamboo casing should be installed in rows. The pattern in which the

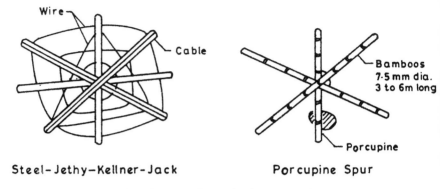

Fig. 9.9 Porcupine made of steel or timber

porcupines should be placed is shown in Fig. 9.7 where the angle of deviation of the row of porcupines with the concave bank is higher than the angle of deflection of the flow or river i.e. ϕ. The length and spacing of porcupines and also its inclination should be best determined from the model tests as well as from the study of prototype itself in each individual cases. The top of the porcupines should be at mid depth of HWL of the river. In between the rows of porcupines block pitching should be made in the concave bank up LWL. Below that the bank should be protected by launching apron, the length of which should cover the anticipated scour depths.

LITERATURE CITED

1. Isbash, S.V: Construction of dams and other structures by dumping stones on to flowing water; Trans. Res. Institute, Hydraulic, 1935, Leningrad, 17, p 1266
2. Indian Standard IS 8408: Criterion for river training works for barrages and weirs in alluvium (1976).
3. River Behaviour Management and Training. Central Board of Irrigation and Power, New Delhi, Publ. no. 204. (1989)
4. Lacey, G. (1930) Stable channels in alluvium, Proc. Institution of Civil Engineers, London, Vol. 229.
5. Ghosh, S.P, Bhandari, P.C., Roy, S.C., and Roy, S.K: On some aspects of bank stability of tidal channel, Proc. Seminar on Bank Erosion and Protection, River Research Institute, West Bengal, 1990.

CHAPTER 10

ENGINEERING ASPECTS OF DREDGING

10.1 INTRODUCTION

Dredging can be termed as the underwater operation with the help of various equipments for removal of materials. Dredging operation in India is presently mostly confined for the maintenance of shipping channel, desilting of harbour, lock entrances and a separate organisation exists for the purpose, namely the Dredging Corporation of India, namely DCI, with Headquarters at Visakhapatnam under the Ministry of Shipping and Transport, Government of India. Dredging is also resorted to for desilting of drainage channels, lakes and reservoirs so as to make them more functional.

Dredging is a high capital incentive and considerable investment is needed in dredging equipment which varies from million to billion dollars depending on sizes of equipment. Dredging is really a far-reaching and extremely important activity. The economic development of many countries is dependent on the infrastructure created as almost all the ports depend on dredging for their sensitivity.

The necessity of dredging arises when low-lying country has to be raised for protecting land from seas, for accesssibility of rivers to navigation. Initial dredging required to open up shipping channel due to inadequate availability of depths or in a harbour site having insufficient water depths is known as capital dredging. Subsequent removal of material for keeping the shipping channel open is known as maintenance dredging.

10.1.1 Capital Dredging

The dredging required to improve upon the existing depth to allow the access of large ships or for the extraction of materials for reclamation. Generally these works are carried out without interruptions of port operations according to suitable timeframe drawn by the developmental activities of the port.

10.1.2 Maintenance Dredging

Maintenance dredging is, however, necessary for the day-to-day existence of the port which cannot be delayed. Depending on dredger used deepening work constitute a serious obstacle to navigation. The execution of dredging operation will be different depending on whether siltation occurs in a more or less uniform way or appears within a short period of time.

In location where siltation occurs in a more or less uniform way, it will be possible to remove siltation almost parallel to their appearance with a dredger with an annual capacity equal to the annual quantities of silt to be removed. Where siltation occurs, in a short time period and the desired water depth is seriously affected it is necessary to remove the deposits within shortest possible time by means of a dredger with a capacity much larger than that required to dredge the same amount over a longer time.

To summarise the aim of dredging operations are:

(1) To maintain and deepen the existing navigable waterways and harbours and the approaches to them, inland as well as offshore (maintenance and capital dredging)

(2) To develop and create ports and harbours, basins, canals, channels, marinas and other waterways.

(3) To improve or to maintain river discharge capabilities by deepening natural water depths or by watercourse realignment for such purposes as improving water qualities, flood control and navigational facilities (partly maintenance and partly capital dredging).

(4) To provide landfills to raise the level of low lands and to reclaim swampy and marshy lands so creating newland areas for the purpose of agriculture, industries, habitation and improvement of environmental conditions.

(5) To digout trenches through areas with bad soil conditions and to provide landfills for the construction of roads, dams, causeways and foundation engineering works.

(6) To extend land areas into the sea and to construct artificial islands for various purposes.

(7) To construct control or protective structures for waterways and coastal areas such as dikes, levees, groins, jetties and breakwaters.

(8) To provide fill materials for beach protection and nourishment including protective dune construction.

(9) To excavate underwater areas for foundations for construction purposes and underwater trenches for pipelines and cables and to provide fill materials to cover up trenches for the protection of pipelines.

(10) To recover subaqueous deposits like sand gravel shell and clay for construction purposes and furthermore minerals for precious metals and fertilizers i.e. wet mining operations.

10.1.3 Sedimentation in Navigational Channel Subjected to Tidal Action

The tidal currents move sediment along alluvial river beds with loose bed material of various grain sizes. The sediment transport is considerable. Because of the periodic reversal of the direction of flow and the fact that flow through the deep shipping channel which is relatively narrow compared to the overall width of the river is not always parallel to the main axis, there can be heavy sedimentation in the shipping channel. Thus continuous maintenance of dredging is required. The siltation thickness in the navigational channel can be assessed from the following relationship[7] considering three hours as the one unit of time.

$$P = 295k \, \frac{\gamma_s \omega}{\gamma_0} \left[1 - \left(\frac{H_1}{H_2} \right)^3 \right] \left[(V_1 + V_2)^2 \, \frac{\sin \theta}{gH_1} \right] \qquad (10.1)$$

where, P = siltation thickness on a unit area at one unit of time (m/time. m^2); V_1 = composite velocity of wind-driven and tidal current; V_2 = mean oscillating velocity of water particles = 0.2 Ch/H where C, h and H represent wave celerity, wave height and water depth respectively; H_1, H_2 = water depths on shoal and dredging depth of navigational channel respectively; θ = angle between V_1 and axis of the channel; ω = mean settling velocity of suspended sediment; γ_s, γ_0-specific weight of silt particles and dry volume weight of deposit respectively; k = 0.4 coefficient.

One can convert the above relationship to estimate annual siltation thickness. The other method of estimating siltation thickness is by survey over different period of time. The average depth of siltation thickness per section of certain dimension derived from the soundings can be expressed in suitable timeframe.

10.2 DREDGE EQUIPMENT

The various types of earlier equipments (antique) used for dredging are as follows. Leonardo de Vinci invented a dredging wheel having catamaran hull with the possibility to readjust to the dredging depth. There is, however, no evidence that Vinci's design was ever tried out in practice. A century later dredger builder, Verantius suggested that a comparable piece of equipment could be used for agitation dredging. Another century later a report on wheel dredgers appeared in French. Dutch mud mills were developed over a period of more than two centuries. The horse-powered Amsterdam mud mill was used as a tool which could efficiently deal with the problems encountered in dredging ports and channels. Later on they are replaced by steam-driven steel dredgers. Besson in the year 1565 published the design of endless bucket chain. The system was used by an Italian Engineer for the cleaning of rivers and canals in France. The first steam-driven dredger was built in the latter part of the 18th century which has the necessary combination of elements for a practical design. In the year 1847, a self-propelled steam dredge boat known as Lavaca was built in USA in Lousiville, Kentucky. The vessel was fitted with real dredge ladders and worked with four mud screws. The success of modern dredging is to a large extent the result of the development of the centrifugal pump. This led to the design of a new set of equipments and changed the dredging scene dramatically, since it made possible to deal with situation on scale unseen so far.

Later, innovation of dredge equipments were provided with automation facilities in the engineroom on the deck and on the bridge in order to save expensive labour. Lots of instrumentation was also provided in pipelines and in the hopper compartments for measurement of flow rate, concentration and degree of loading. This reduces the necessity of carryingout extensive surveys and calculations. Further, electronic positioning systems navigational aids and instruments are also introduced. As for example, by decreasing positional margins by 1 m the amount to be dredged for the approach channel of the port of Rotterdam would decrease by $1 Mm^3$ or six weeks work of a medium sized trailer hopper dredger.

In view of environmental restrictions research into the science of dredging and disposal management techniques has entered a new era. The mechanical engineers and shipbuilders in collaboration with dredging companies have studied the problems of decreasing dredge-induced turbidity, resuspension and dilution of the silt, precision dumping, prevention of leakage, watertight clampshells and submerged diffusers.

Civil Engineers have explored the construction of safe disposal sites, artificial islands, capping operations and impermeable linings. The marine biologists have investigated the effects on the banthic fauna, bio-availability, bio-accumulation and characteristics of the biota. The chemical engineers have studied the prevention of pollution and the cleaning of the dredged material to acceptable limits.

10.3 TYPES OF DREDGERS

The dredgers can be classified as mechanical and hydraulic type. Mechanical dredgers function by physically lifting the excavated material from the river bed. They can be either bucket, carb or Draper types. The bucket-type dredgers has a chain conveyor with buckets attached that dump into a belt conveyor. They can be used where depth is not large and are specially suited for trenching under water. And endless buckets chain is driven continuously from a tap tumbler around a ladder formed in the middle of the floating vessel. The ladder can be lowered or raised by the line and the chain of buckets is operated by a big wheel, the digging is done by the buckets passing at the bottom. The dug material is brought upto as the chain moves round and can be dumped in the hopper of the dredger itself.

The material can also be transferred to the barges alongside by means of a chute. The buckets are provided with manganese steel lips for hard wearing. Electric motors are used to drive these dredgers and they are capable of dredging to a depth of 15 to 20 m under water and each bucket can handle upto 225 litres beginning from 85. The elevator speed may vary from 15 to 20 mpm. This type is suitable for soft or fractured rocks. A 85 litre bucket conveyor may give an output of about 75 cm/hr. The grab or clamshell dredger are used in case the excavated material is rock or hard material requiring blasting and also at places where the spoil area is at a distance from the excavation area and requires hauling. It essentially consists of a grab suspended by a cable or chain from a crane or an extending boom. It can raise, lower and close a bucket and swing it on a boom. The grab on lowering excavates the under water material. It is then hoisted to the surface and the load is dumped into the hopper. The grab cranes are operated by steam, diesel or diesel-operated electric engines. Normally, a central diesel electric plant supplies power to various grab cranes, wrenches and for propulsion. The number of grab cranes mounted on a dredger may vary from 1 to 4. The machine bucket resembles to a clam which is a shell fish with a hinged double shell. The front end of the machine is crane boom with a specially designed bucket. The bucket is generally comprised of shells which are hinged at the top and are provided with sharp edge or teeth at the lower ends. The grabs and their cutting edges are available in various shapes for use to suit various materials. The bucket selection depends on the specific job requirements, a heavy bucket with sharp edge is required for hard soils whereas a light bucket with plane edge is suitable for soft materials. It is also useful in lifting heavy rock pieces. The operation consists in lowering the bucket with shells open over the material to be dug till a good contact is made with it which is then closed in through the clogging line. As the two shells close in the weight of the bucket enables it to dig into the material thereby filling it. It is hoisted later on and swung to the position of dumping.

The dipper dredger is similar to that with land-type Crawler shovels. It is mounted on a floating vessel and then stabilized through a pair of spuds held firmly to the bed. It carries an inclined frame in the bow to hold the boom guy wires. A dipper stick runs through the middle of the boom and is worked by a rock pinion arrangement. A dipper bucket with a flap is attached at the end of the dipper stick, and is moved up or down with the help of hoist cable. The boom is capable of swinging through an angle of 180 degree which enables it to deposit the material on the bank or on a floating barge. They are capable of digging up to a depth of 15 to 20 m under water, and a maximum dumping range of about 35 m. They can employ bucket sizes upto 12.25 m^3 with a production rate of about 50 m^3/hr/cm of bucket capacity and can operate easily in confined spaces around docks and narrow channels and further is capable of excavating in hard soil boulder beds and rocks that breaks into larger pieces. The hydraulic type of dredgers are the ones mostly used by the industry. In this type the dredger moves the soil by suction and pumps through propelling. They are available in many sizes and is determined from the amount of material to be removed or desired level of production.

In this category one can have either a cutter suction dredger or a trailing hopper dredger. In the former the suction pipe carries at its lower end a cutter having a universal joint at the top. The type of material to be excavated decides whether a cutter head will be required at the end of the suction line. A cutter head is used when material has to be loosened and cut up into small enough pieces to get into the suction line and flow through the pipes to discharge. In the case of freeflow material like sand and gravel there is no need for a cutter head. The dredgers of this type are of length varying between 45 to 140 m in length and 9 to 30 m in width and they dig up depth varying from 4.5 to 15 m and normally driven by diesel electric power. In the case of trailer hopper suction dredger the material to be excavated is loosened by a high pressure water jet and the loose material is then sucked in the suction pipe and finally is discharged through the discharge pipe. The maximum length of pipe line that can be used depends on the specific gravity of the material to be pumped. Sometimes a booster pump is installed in discharge line if velocities get too low. The abrasive material, velocity and specific gravity of pumped material cause the wear on the discharge line. Figure 10.1 show a cutter suction and trailer hopper suction dredger.

Dredging equipments are very costly, a cutter section dredger and trailer hopper suction dredger of size 8500 hp and 5000 m^3 capacity can be anything like 60 million US $ with an economical life span of 15 yrs Owning a dredger implies foreign exchange expenditure irrespective of the

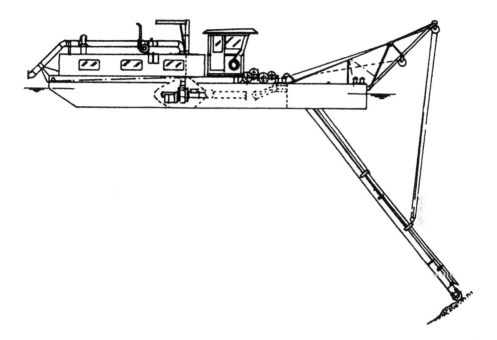

Fig. 10.1 A cutter suction and trailer hopper suction dredger

fact whether the dredger is in operation or not. Since the technical depreciation of the equipment often takes place within a much shorter time span than its economic depreciation, a high rate of occupancy of the equipment is required to avoid unnecessary subsidies.

10.4 DREDGE SPOIL AND DREDGE DISPOSAL

10.4.1 Dredge Spoil

Shipping channels in estuaries and at the entrance to ports may require frequent dredging to keep them open and the dredged material is barged out to sea and dumped. This dredging spoil particularly from industrialised estuaries, may contain appreciable quantities of heavy metals and other contaminants which are then transferred to the dumping grounds. Ports, harbours, rivers and approach channels often need regular dredging to keep them open to shipping. The dredged material may be used in land reclamation, but is more often dumped at sea. In England and Wales alone, about 28 million tons of dredging spoil is dumped annually at 60 licensed offshore sites. The composition of the spoil is varied; for routine dredging

144

it is usually fine silt, but if new channels are constructed or old ones deepened, it may consist of boulder clay or chalk, etc. Unlike other solid wastes dumped at sea, the spoil cannot usually be distinguished from the surrounding substratum at the dumping ground and depending on the bottom currents, it may not stay where it was dumped, but gradually be transported to other areas, including the place it originally came from. Dredging spoil is often anoxic and contaminated with metals and persistent oils. Pollutants accumulating in estuarine sediments are transferred elsewhere in the course of dredging. An idea of quantities of metals expressed in kg/day in dredging spoil dumped in North Sea in the year 1981 by different countries is shown in Table 10.1.

Table 10.1

Country	Cadmium	Mercury	Chromium	Copper	Lead	Nickel	Zinc
Belgium	0	31	6,978	1,761	4,422	1,448	2,721
Netherlands	151	32	2,693	1,457	3,789	873	11,486
U.K.	14	19	1,149	804	1,213	390	3,425
Totals	165	82	10,821	4,026	9,425	2.711	42,141

Note: The total includes small quantities contributed by Denmark.

Table 10.1 shows the quantity of heavy metals included in dredging spoil dumped in the North Sea each year by Belgium, Holland and U.K. These totals conceal a wide variation. Spoil from harbours and estuaries with intense shipping activity and heavy industry is much more contaminated than that from other areas. Dredging spoil from Manchester Shipping canal for example contains, 20.7 ppm of mercury and 5080 ppm of lead. The Tees estuary spoil contains 7 ppm of mercury and 3000 ppm of zinc but only 320–460 ppm lead. Swansea dock's spoil contains a high concentration of cadmium i.e., 18 ppm. At the other extreme the chromium concentration in the spoil from Newhaven is 1 ppm compared with 1500 ppm in one sample from the Tees estuary.

The immediate impact of dumping dredging spoil is smothering of the benthic fauna in the dumping grounds, though some animals are able to burrow the surface of the dumped material without much difficulty. A study was carried out in the fjord at Uddevalla on the west coast of Sweden where dredging was going on in an area where surveys of benthic fauna had been made in the year 1971-72. Immediately after the dredging of a new channel there was a loss of diversity of stations near the site of operations but the position was restored a year later. Benthic animals in the area accumulated high concentrations of metals when dredging was in progress, but returned to former levels within a period of 18 months.

10.4.2 Dredge Disposal

Dredge disposal, primarily arising out of maintenance dredging is very crucial as is often linked up with contamination. The contaminants normally found in dredged material generally come from industry based near the upper reaches of the river in question. There exists no exact figures of worldwide contamination as yet to the writer knowledge. It is, however, estimated that some 90 percent of the total of the material to be disposed of is clean or at least not permanently not harmful. The remaining amount would require special care and the disposal of the seriously contaminated ones would pose a great problem. It is believed that the cost of temporary upland storage of highly contaminated soil as well as cleaning up is about 10 times more expensive than disposal at sea; disposal of heavily contaminated material could even show a factor of twenty. The disposal sites are respectively (i) sub-sea, (ii) shoreline and upland. Contaminant is affected in a number of ways with special attention being paid to the prevention of leakage. On ground special care is taken to provide impermeable linings while subsea disposal areas are capped with clean material. Four types are generally distingusihed when dealing with contaminated dredged material. These are: (i) Comprising roughly 90 percent or more marine silt and, 50 percent marine silt, light contamination is expected, (iii) comprising more than 90 percent fluvial silt of the relatively contaminated depending on upriver water quality, (iv) sludge from highly contaminated hot spots. Disposal above is affected as follows.

(a) uncontaminated material is disposed of at sea with due precautions as to the sites or is used beneficially, (b) highly contaminated material is disposed of in special contained facilities upland or subsea. The costs of the treatment to produce acceptable top soil/land filling material has not yet approached commercial validity. Highly contaminated material of which toxicity is of great concern is disposed of in tightly controlled upland facilities to prevent mobilisation of the contaminants.

Regarding capping level bottoms it should be remembered that this method of disposal can only be effective for relatively cohesive material which does not spread over two great a surface. Capping should be executed from the edges to the centre to ensure that everything will be covered and that no material will be squeezed from under the cap. The capping material must have a particle size big enough not to be dispersed by currents or waves and small enough to deter sinkage into or mixing with the contaminated dredged material. The dumping and capping vessels must be equipped with a very accurate positioning system to achieve successful capping. Results of the capping operation can be checked with vibro core borings, preferably in conjuction with a seismic survey. Apart from above the dredging method can be adopted to optimise the capping procedure. For example, a bucket dredger when used to excavate the suspended materials will enable them to keep their natural cohesiveness. The sea disposal be carried with a stationary split barge to make the material go down in one lot. Capping with sandy material should be done by a slowly moving vessel equipped with bottom doors permitting a certain spreading of the material as if the capping material go down as one cohesive lump it could penetrate into the layer instead of capping.

The polluted materials should also be released very near to the bottom through a discharge pipe mounted on the discharging vessel and this is possible in shallow dumping areas and for this a trailing suction hopper dredger may be adopted.

As regards the disposal by dilution, physical dispersion is one of the principal means of fighting the impact of waste materials on the marine environment. The greater the dispersion, the quicker the marine environment will neutralise the disturbing agents and quicker it will regain its predumping position. To effect dilution rapidly on a large scale is to choose a suitable unloading vessel. A vessel pumping its spoil water mixture overboard while navigating at full speed will create a better dilution effect. The other factor influencing dispersion is the presence of currents at dumping sites which may be decisive factor for the choice of disposal site. To achieve better dilution some special care measures are required which entail the selection of special disposal sites. In deep oceans the areas specially fit to receive pollutant materials are: (a) hypersaline basins and (b) submarine canyons. However, the sites being far away from dredging site the transport cost of dredged materials will be heavy on the costs of the entire operation. Special disposal sites for highly polluted dredged materials can be created by construction of artificial offshore islands where the waste can be isolated from any contact with the marine environment. Further, the islands can be put to beneficial use by serving as industrial or recreational areas. The seas and oceans offer also the choice of other dump areas depending on biological, sedimentological and dispersional characteristics of the sites. In general the biological characteristics of proposed dump sites are subject to seasonal changes. Sometime it may be necessary to discharge the dredged materials below the zone of maximum plankton productivity i.e., delivering it some 15 m below the transporting vessel. To take care of environmental problem several alternatives are devised such as siltmaster. In this case the soil water mixture is not overflowing into the sea but to silt spaces on both sides of the hopper. The unloading of the silt spaces is done by opening doors between the hopper compartment and the silt spaces. The other is the turbidity-control overflow system (Fig 10.2) Here the whole overflow cylinder is kept full by a regulating discharge valve. This helps the overflowing mixture to be kept free from the air bubbles which in the traditional systems cause turbidity over a wide area behind the dredger. The conical inlet and the discharge under the keel of the ship also help to minimise turbidity. Direct loading from trailing hopper suction dredgers into high capacity ocean going barges combines two advantages, the overflow effects are halted and the production costs connected with long transport distances can be kept at a reasonable level.

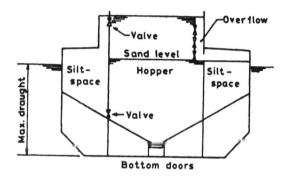

Fig. 10.2 Design of turbidity-control overflow system

Hopper loads

In many countries production is calculated by counting the number of trips. Every trip is assumed equal to a load of the theoretical volume of the hopper which gives unreliable results. A hopper filling ratio of 100 percent is very unlikely. In the case of sand the accuracy of the quantity calculation can be improved by a filling ratio factor which is less than 1. This factor has to be calculated for particular cases by measuring the average load volume of 10 trips. In the case of load of mud or mixed sand/silt load the method does not provide accurate quantity information at all. On average the filling ratio varies from 0.2 to 0.6. A more reliable method of calculation of load is to measure the displacement of the vessel before and after the loading which is normally done by a load recorder. For reliability of the results regular calibration of the recorder is necessary.

A widespread method of quantity calculation on board of the hopper dredger is to measure the volume of the load in the hopper. Generally sand loads do not present much difficulty, however mud or mixed load give ample problems in estimating the quantities of solids in the hopper. After the loading process has been stopped the level of the solid cargo in the hopper is measured on several locations as well on portside as on starboard side with a marked measuring line with half a ball. It is envisaged that due to the specific weight of the half ball i.e., 1.35 kg/litre the ball will be stopped by a density of the layer equal to or more than the specific weight of the ball. The sum of all levels is divided by the number of observations to get the average height of the level of solid load. A list compared officially by a ship surveyor provides the necessary relation between a certain level in the hopper and the comparable volume. Next is to calculate the volume of the fluid mud load. For this samples are taken halfway between the average level of the solid load. The samples are then analysed to find the solid content or fraction. It is common practice to take a sample of approximately one litre at a level halfway between the top of the solid load and top of the fluid mixture. The sampling bottle can be opened remotely at the desired depth. The above implies that the vertical density distribution in the hopper follows a linear increase with depth. As for example consider a hopper dredger with hopper volume of 4000 m^3 with a measured load of sand, mud and water, is measured following the above procedure. The calculation of solid volume can be done as follows.

Say average level of the sand equals H for the example is 2.8 m. The list reading H as shown in the Table 10.2 below shows a volume of 2500 m^3 of sand. So the fluid mud equals 4000–2500 = 1500 m^3

Table 10.2

2.6	2580
2.7	2540
2.8	2500
2.9	2460
3.0	2420
.,	,,

The sample is taken at a level of 1.4 m is centrifuged or after consolidation gives, say 35 % of solids. So the fluid mud contains 525 m^3 of solids. The total load is therefore = 3025 m^3 of solids to be paid for as production figure for the particular trip.

Land disposal

By disposing the dredge spoil on land one can reclaim some areas. Some preparations are, however, necessary for the reclamation of the area. Generally one has to make a layout of the reclaimed area and it has to be properly bunded all along depending on the topography. The elevation of the bund section will have to be properly determined with respect to datum. The bunds are generally erected by either mechanical or manual excavation of the soil from inside the reclamation area with suitable margin kept between the inner edge of embankment upto the excavation point. Generally the bunds facing the waterfront will be constructed higher as most of the deposits settle there because at that location shore pipelines will enter the reclaim areas. For draining the water out from the area suitable outlet arrangement have to be made and the outlet will have to be connected with an existing drainage system such as tidal creek. The outlet system will have to have an adjustable overflow system and depending on the composition of the dredge spoil in the reclaimed soil the overflow will be adjusted.

Next it is necessary to check the stability of the embankment/bunds following soil mechanics principle. For this purpose generally a borehole data is obtained with average soil penetration test values for different layers. Soil penetration values or N-value is the number of blows over 30 cm penetration after a first seating drive. Knowing the N-values other soil parameters such as undrained shear strength of the top layer of soil can be found out. The bund height can then be determined based on the analysis of the critical circle failure plane.

10.5 DREDGING COSTS

Before the dredging contract is awarded it is necessary to estimate the cost of the proposed project. This involves the estimation of many relevant dredging parameters such as capital costs, fuel costs, labour costs, production costs and travel time among other variables. Then it is necessary to identify parameters to which dredging costs are more sensitive. They are then studied to improve dredging project estimating procedures.

The sensitivity of costs to parameters may also be used as tool for evaluating dredging investment decisions. Making use of regression models applied to economic data the sensitivity of dredging cost with respect to various parameters may be revealed. A simple model is shown in the box.

> Cost = A_0 + A_1 × Capital + A_2 × Fuel + A_3 × Labour; where A_0, A_1, A_2, A_3 are the coefficients; cost = unit cost of dredging say (Rs/m); Capital = capital costs (in Rs); Fuel = fuel costs (in Rs); Labour = labour costs (in Rs).

According to econometrics these variables are also dependent on other variables.

As for capital costs they are a function of interest rates, fuel costs a function of fuel reserves, the demand for fuel and the value of rupees in the foreign exchange market; labour costs a function of the availability of labour and industrial capacity. Since these variables are again dependent on other variables, the model becomes more complicated:

$$\text{Capital} = B_0 + B_1 \times \text{interest}$$
$$\text{Fuel} = C_0 + C_1 \times \text{reserves} + C_2 \times \text{demand} + C_3 \times \text{value}$$
$$\text{Labour} = D_0 + D_1 \times \text{Avail} + D_2 \times \text{Indust}$$

where, B_0, B_1, C_0, C_1, C_2, C_3, D_0, D_1, D_2 are coefficients.

Interest = interest rate in percent; Reserves = fuel reserves in India (litres); Demand = fuel demand in litres; Value = value of the rupees in foreign exchange; Avail = availability of labour (no. of workers); Indust = industrial capacity.

With data available for the variables the model can be solved by the least square techniques. Such a solution for example could be

$$\text{Cost} = 1.50 + 5 \times 10^{-6} \times \text{Capital} + 3 \times 10^{-6} \times$$
$$\text{Fuel} + 8 \times 10^{-6} \times \text{Labour}$$

The coefficient may be used to determine the sensitivity of unit costs to the parameters in the model. For example, if the capital costs are increased by Rs. 10,0000, the unit costs will rise by 0.5. If fuel costs are increased by Rs. 100000, the unit costs will rise by Rs. 0.3 and if labour costs are increased by Rs. 100000, the unit costs will rise by 0.8, so here the unit costs are most sensitive to labour costs which means it would be wise to automate the dredge and if fuel cost is highest then fuel efficient machinery may be thought of and if capital cost is highest a policy of increasing labour intensity should be preserved.

In various countries escalation clauses of the following general form have been used for dredging contracts.

$$P = p \left(a \frac{W}{w} + b \frac{M}{m} + c \frac{T}{t} + d \frac{E}{e} + x \right) \qquad \dots \quad (10.2)$$

where, P represents new price; p = old pirce; W, M, T, and E = various cost categories at the redetermined level; w, m, t, and e = the same elements at their earlier level, which will those be prevailing on the date of tendering in those cases where cost escalation is calculated on termination of the work. In other situations the earlier price levels will be those which prevailed last year, last quarter or month depending on how often the cost escalation is made up. The coefficients a, b, c, d and x indicate their proportions in the total cost price and when added equals unity, i.e. $(a + b + c + d + x) = 1$.

10.5.1 Radioactive Tracer Study

The use of radio active tracer techniques help in minimising the cost of dredging by reducing the distance to the dumping site. A balance must be established between the cost of transpsort and the eventual return of spoil to the deepened areas, or navigational channels. The cost of transport is known but the efficiency of the dumping site is very difficult to establish. For the latter it is necessary to study the transport of spoils in suspension during dumping and the transport or erosion from the bottom of the spoils by current waves. The transport of suspended sediments is charactersised by particle direction and velocity, dispersion coefficients and siltation rate. To ascertain these radioactive tracer techniques can be used to follow the actual movement of sediments so as to furnish information about transport parameters in practice.

A tracer is a radio active chemical which is added to the dredging spoil so as to label the sediment. Measurement of the radio activity during movement of the sediment provides information about the distribution and the velocity of transport of the particles. For sediments with a mean diameter below 300 mm one kilogram of labelled sediment is sufficient to estimate the transport parameters.

Automatic or semi-automatic systems of injection are used to reduce the handling of radioactive material to a minimum. The radio activity is detected by highly sensitive counting instruments and the amount measured is proportional to the mass of tracer particles present around the probe. Using a calibration (laboratory) of the instrument, the counting rate can be transformed to a sediment concentration and the distribution of counting rate is representative of the distribution of sediment in suspension.

The choice of radio-isotope depends on the length of time planned in the experiment. Normally the injection is done during maximum flood and ebb velocities and lasts several hours during which the distance travelled by the cloud of sediment in suspension is measured. Figure 10.3 shows a typical longitudinal recordings along the cloud together with time and space positions of the maximum measuring counting rates. The expression for maximum counting rate measured at various times (t) is as follows

$$R_{d_{max}}(t) = \Sigma F_n(t - t_0)^{n-1} \qquad \dots \text{(10.3)}$$

$$F_n = \frac{EA_0}{4\pi K_d \sqrt{D_x D_y}} \left(\frac{-1}{n}\right)^n \int_v F_v \left(\frac{V}{H\phi}\right)^n dV \qquad \dots \text{(10.4)}$$

$$F_0 = \frac{EA_0}{4\pi K_d \sqrt{D_x D_y}} \qquad \dots \text{(10.5)}$$

$$F_1 = -F_0 \int_v F_v \left(\frac{V}{H\phi}\right)^n dV \qquad \dots \text{(10.6)}$$

where A_0 = initial tracer activity, E = calibration coefficient, T = time at which instantaneous injection of total mass, taken place, $H\phi$ = total amount

152

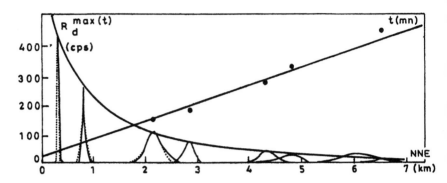

Fig. 10.3 Redioactive tracer data

of sediment in suspension over the depth, $F_v dV$ = summation for all size fraction, K_d = water height at a particular depth, n = time interval of measured $R_{d_{max}}$.

Equation (10.3) can be matched to actual field data, providing F_0, F_1, etc. until no more improvement in matching is obtained. In practice, F_0 and F_1 suffice for the analysis of very fine sediments since during the flocculation process all particles have about the same unknown field settling velocity V, which is equal to α times the velocity measured in the laboratory, i.e., $V = \alpha\, V_{lab}$ which provides

$$\sqrt{D_x\, D_y} = \frac{EA_0}{4\pi\, k_d F_0} \qquad\qquad \dots\ (10.7)$$

and

$$\alpha = \frac{F_1\, H\phi}{F_0 V_{lab}} \qquad\qquad \dots\ (10.8)$$

Another relation connecting α is as follows:

$$\alpha = \frac{mVkg}{CV_{lab}} \qquad\qquad \dots\ (10.9)$$

where C – Chezy's coefficient; g – acceleration due to gravity; V – settling velocity; k – Karman's constant; $m - V/\sqrt{\tau_0/\rho}$; τ_0 – shear stress and ρ – density.

The transformation gives a system of two equations with two unknowns α and ϕ which can be solved by a graphical; or numerical method. The equation (10.7) gives the geometric mean of the dispersion coefficients while (10.8) gives the ratio between the field and laboratory settling velocities. Individual estimates are obtained from activity readings along and across the radioactive cloud by matching the recording to the equation

$$R = R_{max} \exp \left(\frac{-l^2}{4D\tau_0} \right) \qquad \dots (10.10)$$

where, $l = 0$, $\tau_0 = 0$ are obtained at $R = R_{max}$
The settling rate S is given by

$$S = -100 \, \frac{F_1}{F_0} \text{ (in percent)} \qquad \dots (10.11)$$

The transport rate is directly obtained from the time and position of the measured $R_{d_{max}}$. Thus it is possible to obtain the siltation rate, dispersion coefficients D_x and D_y, the longitudinal transport velocity, an estimate of the ratio between actual field and laboratory settling velocity and the proportion of sediment in suspensions to the amount which was immediately deposited during the dumping of spoils.

LITERATURE CITED

1. Riddell. (1983). Dredging; the Need for Civil Engineering Education, Terra Et Aqua, Nr. 26, IADC, Duinweg, 21 Hague.
2. Vlieger de H. (1983). Implications of the London Dumping Convention for a dredging reactor, Terra Et Aqua, Nr. 26 IADC, Duinweg 21, Hague.
3. IAPH (Ad-Hoc Dredging Committee) Special care measures for safe disposal of dredged material in the marine environment, IAPH paper submitted to the IMCO Scientific Group Meeting, May, 1981.
4. Sauzay (1968). Appraisal of radioactive tracer techniques in dredging operations, Terra Et Aqua Nr. 10 IADC Duinweg 21, Hague.
5. ICRP Report of Committee II on permissible dose for internal radiation, Pergamon Press, Oxford (1959).
6. Crickmore, H.J., and Lean, G.H. The measurement of sand transport by means of radioactive tracers, Proc. Royal. Society of London, Series A, Vol. 266, 1962.
7. Second International Symposium on River Sedimentation Nanjing, China, Oct 11–16, 1983.

CHAPTER 11

POWER DEVELOPMENT FROM TIDES

11.1 INTRODUCTION

The importance of development and utilisation of alternate sources of energy needs hardly to be emphasised. The search is therefore continuing for alternate sources of energy. Any development for extraction of energy from the gift of nature has to ensure that the natural balance or ecological system is not disturbed beyond repair. In other words, systems which produce pollution and hazardous effects on the environment must be avoided. Tidal energy being perennial, non-pollution and hazard-free in nature, offers good scope for development. The tidal effect is caused mainly by the mutual attraction of the moon and the earth when rotating in an almost monthly period around an axis through their common CG and oriented normal to the plane of the lunar orbit. Along an earth diameter parallel to the common axis the centrifugal force induced by rotation around the axis and the lunar gravitational attraction are balanced. The resultant force on a fluid at the ends of this diameter points towards the earth's centre and so the fluid element at these stations remains with its surface parallel to the earths surface. For a fluid element on the side away from the moon the centrifugal force due to rotation around the common axis is stronger than lunar gravitational attractions; the mean surface there deviates slightly from the nearly spherical original form, creating a hump oriented away from the moon. On the opposite side the stronger gravitational influence of the moon forms a hump pointing towards it (Fig. 11.1).

A similar interaction exists between the earth and the sun. For the same effects two flood humps are generated at the ocean in the direction towards the sun and away from it. They are naturally smaller than the two due to interaction of the earth and the moon along a straight line during new moon and full moon. The highest tidal range, designated spring tide, occurs then. When the earth is in the corner of a rectangle formed by the sun, moon and earth, the two humps have to be subtracted from each other. The lowest tidal range, i.e., neap tide, occurs then. Tidal power plants harness the tidal range for generation of power. However, the water head available will be small. The height of tides varies from about 1 m to 15 m according to the geography of the coast, location of the estuary, depth of water, direction and speed of the prevailing wind, etc.

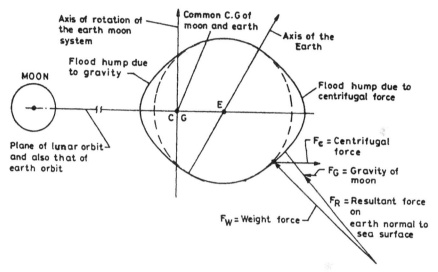

Fig. 11.1 Effect of moon-earth gravitational attraction on tidal activity

11.2 HARNESSING TIDAL POWER

The rise and fall of tidal water levels provide a potential head that can be used for the generation of hydropower.

The engineering solution to recovery of energy from the tides is construction of a barrage across the estuary with sluices. The sluices help the basin to fill during rising tide. They are closed at high water and during ebb tide the water is released through the turbine. In this scheme a number of daunting problems have to be faced, namely:

i) the production of electricity will vary according to the phases of the moon;

ii) the solar and lunar cycle do not synchronise and our habits and way of life are generally based on solar aspects, as is the demand for electricity;

iii) output will not be in phase with the requirments, capital investment will be large and so also the periods of little or no output.

156

Depending on the particular situation, different types of barrage regulations are required to fit in the particular development, to it:

(i)   single basin type, generation of ebb tide only;
(ii) single basin, generation occurring on the flow of the tide in both directions;
(iii) single basin with two-way generation aided by pump storage;
(iv) double basins with one of the above scenarios as well as a double basin with a pump storage scheme and tidal boost.

The barrage embodies a number of sluice gates and low head turbine sets. Two typical tidal schemes with constant geometric openings at intake and flow-off are shown in Figure 11.2. The single-basin type may have one flow direction which guarantees, the highest efficiency. It has the disadvantage of limited working period or the requirement of high storage volume. The Rance Power Plant (Fig. 11.3) developed by the French firm Neyric has adopted an axial bulb turbine that operates with one basin in both flow directions, i.e., as a turbine and as a pump.

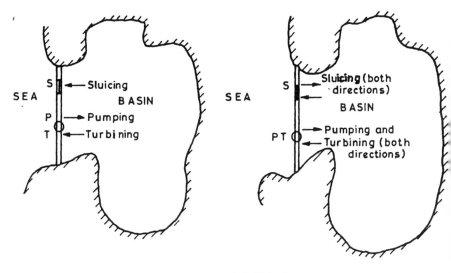

Fig. 11.2 Typical tidal schemes

11.3   OPERATION SEQUENCES

The time sequence of various operation modes in a tidal power plant is shown in Figure 11.4. Owing to the tides the ocean level indicated as O moves nearly harmoniously with respect to time. At its highest level it is connected with the basin by a sluice, which is then opened. When the ocean level starts to fall the sluice gates are closed. As soon as a reasonable difference in height of water levels in the basin and the ocean occurs, the

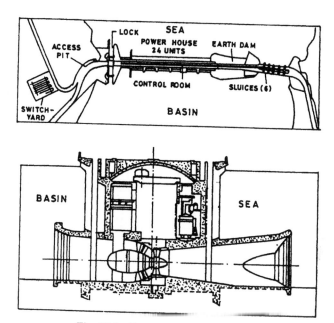

Fig. 11.3 Rance power plant (France)

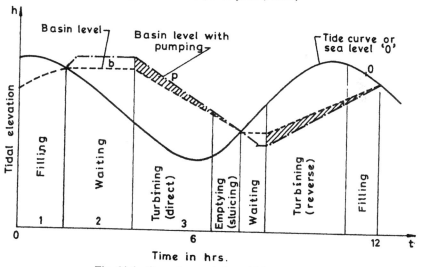

Fig. 11.4 Operating mode in a tidal power plant

set starts turbining from the basin to the ocean. The set is shut down before the occurrence of ebb. The sluice gates are then opened to equalise the level of the ocean and the basin. When a reasonable head between the rising level of the ocean and that of the basin for the second cycle of power generation has been attained, the machine begins to operate as a turbine with reversed flow direction from the ocean to the basin.

158

The best efficiency can be achieved only when the flow direction is retained. This can be achieved by a single-action working cycle. The sequence of operation modes in a tidal power plant with one basin and a single-action effect are respectively (i) filling, (ii) waiting plus eventual pumping and (iii) turbining. Figure 11.5 shows a typical scheme of tidal power plant and cycle of operation.

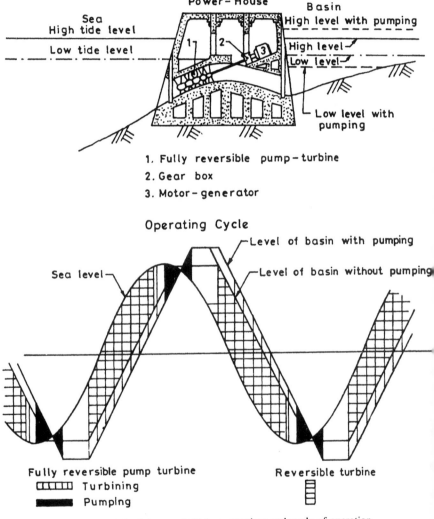

1. Fully reversible pump-turbine
2. Gear box
3. Motor-generator

Operating Cycle

Fully reversible pump turbine
[TITITT] Turbining
■■■ Pumping

Reversible turbine

Fig. 11.5 Scheme of tidal power plant and cycle of operation

The double-basin single effect incorporates two parts, one upper and one lower. The powerhouse is situated between them. Filling gates are built in-between the sea and the upper pool and the sea. This setup permits continuous power generation proportional in output to the difference in heads between the pools. The direction of flow is always from the upper to the lower pool. The flood tide fills the former daily while the latter is emptied by the action of the ebb tide when the emptying gates are opened. This system, albeit costing more, has the advantage of a continuous though limited supply. A typical layout of such a basin is shown in Figure 11.6.

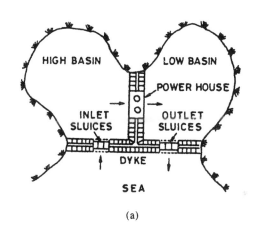

(a)

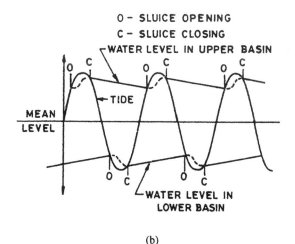

(b)

Fig. 11.6 (a) Co-operating double-basin system; (b) Operation of two co-operating basins

11.4 ESTIMATION OF POWER POTENTIAL

To find out the power potential a hydrographic survey of the area must be made. This will furnish information regarding the cross-sectional area of flow at various levels of tidal fluctuation. Further, one has to collect information regarding the variation of tidal height as well as velocity. A typical measurement of tidal information is shown in Fig. 11.7.

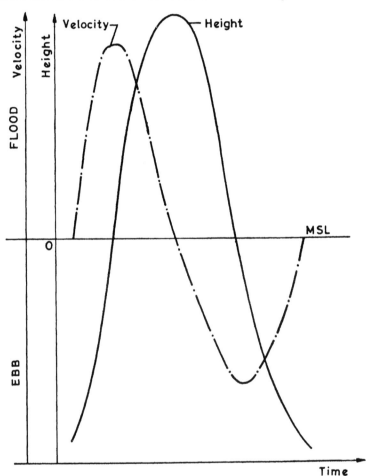

Fig. 11.7 Typical measurement of tidal information

Let Q equal area multiplied by velocity (max) = AV, where A = area a M.S.L, V = maximum velocity during flood or ebb, ξ_0 = maximum leve (range) during flood or ebb tide; δ = phase shift; ϕ = phase angle.

$$(2\delta\pi/T) = [\delta\pi/T/2]$$

where, T-periodic time; $T/2$-half period time;

N = average power = $Q_0\, \rho_s\, g\, [\xi_0/2]$ sin ϕ where ρ_s = density of sea water, g = acceleration due to gravity. Suppose from the survey data, δ = 1h 48 min = 1.8 h, $T/2$ = 6.2 h; ϕ = (2 × 1.8) /6.2 = 52° 15′ 29″; sin ϕ = 0.79078: ξ_0 = 137.00 cm; Q_0 = 6.65 × 10^9 cm³/s; ρ_s = 1.025 gm/cm³; g = 980 cm/ s²; $\rho_s\, g$ = 1.0045 × 10 gm/cm² s². Then:

$$N = 6.654 \times 10^9 \times 1.0045 \times 10^3 \times 137.16 \sin 52° 15′ 29″$$
$$= 7.2495 \times 10^{14} \text{ ergs/s } [1 \text{ MW} = 10^{13} \text{ ergs/s}] = 72.5 \text{ MW.}$$

During ebb tide, Q_0 = –4.4 × 10^9 cm³/s [maximum seaward discharge];

$$\xi_0/2 = -252.984/2 = -126.5 \text{ cm}; (\rho_s\, g) = 1.0045 \times 10^3 \text{ gm/cm}^2\text{s}^2;$$
$$\delta = 3 \text{ h}3 \text{ min} = 3.05 \text{ h}, \ T/2 = 6.5 \text{ h};$$

$$\phi = [(180 \times 3.05)]/6.5 = 84° 27′; \sin \phi = 0.99.$$

Average power $N = [(-) 4.39 \times 10^9 \times (+ 26.5) \times 1.0045 \times 10^3 \sin 84° 27′$ 41″ ergs/s = 5.50 × 10^{14} ergs/s = 55.5 MW.

Hence during ebb tide the average power generation will be 55.5 *MW*.

11.5   TIDAL POWER MACHINES

The low head due to tides predetermines the type of axial turbine to be used, such as the Bulb Turbine or Straflo Turbine. Figure 11.8 shows a cutaway view of a bulb turbine for harnessing tidal power. For the case of double-action plants with one basin, harnessing tidal power is done by a machine capable of direct and reversed turbining. Further, such machines are also required for carrying out direct and reversed pumping. It is therefore necessary to consider the special features of such an axial tubular pump-turbine machine for direct and reversed flow in both the modes of operation.

Consider a bulb-type turbine as shown in Fig. 11.3. Here the flow enters the runner from the side of the bulb with a whirl produced in the stay and mainly in the guide vanes of a conical distributor. In this case of so-called direct flow whilst turbining the draft tube operates as a diffuser. Obviously this works at a higher efficiency as a turbining than under reversed flow when the gate operates as a diffuser. This direct turbining is usually combined with a flow from the reservoir to the ocean. In case the basin is situated at the mouth of a river, larger work is done along with high efficiency. In the case of reversed turbining the inlet whirl of the rotor vanishes since there are no vanes in the draft tube. The channels between the guide vanes operate as a diffuser with modest pressure recovery. As for the design of the gates it is important that they close in case of emergency. To obtain the best efficiency the direction of flow is retained. This can be done by a single-action working cycle. The working principle of a single-action turbine is as shown in Fig. 11.4, namely; waiting plus eventual pumping, filling, turbining, emptying, waiting, turbining, filling, vis-a-vis sea level, basin level and basin level with pumping.

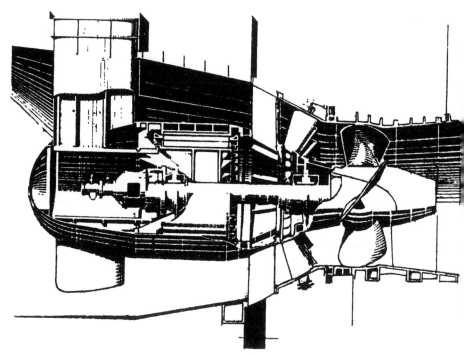

Fig. 11.8 Cutaway view of bulb turbine[5]

The layout of a typical bulb turbine and space requirement for civil engineering works is shown in figure 11.9 both in plan and in elevation, where D stands for runner diameter. The design features of axial turbines are shown in Table 11.1.

Table 11.1 Design features of axial turbines (after Raabe [5])

n_s	n_q	n_q^1	H_{max} (m)	z	$N = D_H/D$	L/t_e	L/t_i	σ	K_u	Kc_m	b_3/D	H/H_2
500	145	2.7	30	6	0.5	1.0	1.7	0.52	1.4	0.44	0.39	0.37
600	174	3.28	15	5	0.44	0.9	1.6	0.70	1.6	0.50	0.40	0.33
700	203	3.84	8	4	0.44	0.8	1.4	0.95	1.7	0.54	0 42	0.28
1100	318	6.0	4	3	0.39	0.7	1.1	2.1	2.3	0.70	0.45	0.23

where n_s = specific speed, $= \dfrac{n\,(\text{rpm})\,P^{1/2}\,(\text{kw})}{H^{3/4}\,(\text{m})}$;

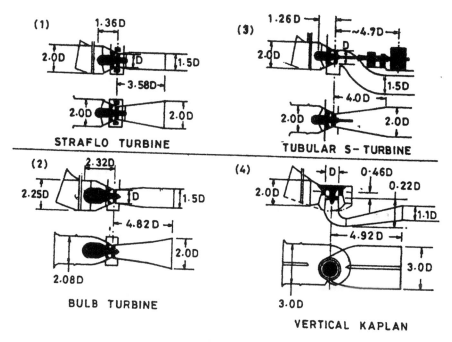

Fig. 11.9

After H. Miller, 1978. Choice of hydroelectric equipment for tidal energy. Symp., Seoul, Korea

$$n_q = \text{specific speed}, = \frac{n(\text{rpm})Q^{1/2} \ (\text{m}^3/\text{s})}{H^{3/4} \ (\text{m})}$$

n'_q = non-dimensional specific speed (type number) = $\omega \ Q^{1/2}/(gH)^{3/4}$;

z = vane number;

L = cord length;

t_1, t_e = pitch at inlet and exit;

σ = cavitation index;

K_u = coefficient of blade slip speed;

Kc_m = coefficient of mean meridional speed;

b_3 = gate span;

H_2 = distance of runner vane axis from bottom of guide vane channel;

D = rotor diameter.

Propeller-type turbines are employed to match the conditions of operation under low heads. That which provides the maximum operating efficiency is the one with a horizontal shaft bulb with the generator components installed in a steel bulb surrounded by the water turbine passages. This type was installed at the Rance tidal plant (see Fig. 11.3). It is reversible, has adjustable wicket gates and adjustable runner blades, and is considered to be highly efficient. Being a multipurpose unit, it is capable of acting as a generator, can be used for pumping and can serve the role of sluices in either direction of flow. Thus this type of unit is capable of operating in six separate roles. It may be mentioned here that reversible units are not essential for a double-basin scheme or a single-basin scheme where the generating flow is always one way; so for the sake of economy fixed blades and non-reversibility should be introduced at the design stage. Bulb-type generators are housed in watertight steel compartments situated completely within the turbine water passages. The minimum possible size must match optimum power output for that particular site.

11.6 DESIGN AND CONSTRUCTION OF TIDAL POWER STRUCTURES

They are gravity structures that rely for stability on their weight after installation. For combined durability, safety and economy, high-strength concrete, whether reinforced or incorporating prestressed concrete structures, has been both tested and proven suitable. The durability of concrete is all important, particularly as the structures will be open to the sea, placing the reinforcements at risk. This aspect led to the development of a special cement, one resistant to attack by marine water, the cement mixture should contain good-quality aggregates and a minimum cement content, the cement/water ratio is vitally important.

The Rance barrage was constructed by the usual methods. The main river barrier consists of sand and gravel fill protected by massive concrete blocks on the slopes. The locks, pump stations and turbine houses were built *in situ* within cofferdams, which amount to some 30 per cent of the total civil engineering works.

Today the construction technic involves concrete caissons, constructed in sheltered conditions and floated into position. The caissons have to be designed for a penetration depth of about 5 m to ensure stability. Where turbines have to be incorporated the width should depend on the type of turbine selected, the normal width being 10 times that of the turbine diameter. Caisson emplacement depends on the conditions likely to be encountered at the site, the nature of the sea-bed, the tides, currents and weather. Advantage must be taken of the weather and tidal conditions. Generally tugs are capable of controlling and placing the largest caissons within 1 m of the selected spot. When flooded down and filled with ballast, penetration of the caissons should occur. Where necessary, joints should be made between the vents and anchoring to the sea-bed done.

LITERATURE CITED

1. Shaw, T.L. (1978). The status of tidal power. *Water Power* 30 (6): 29-34.
2. Sharma, H. R. (1982). India embarks on tidal power. *Water Power* 34 (6): 32.
3. Subrahamanyam, K.S. (1978). Tidal power in India. *Water Power* 30 (6): 42-44.
4. Benn, T. (1978). Tidal power problems and benefits. *Water Power* 30 (6): 27.
5. Raabe, J. (1985). *Hydropower.* VDI-Verlag, Dusseldorf.
6. Simeons, Charles (1980). *Hydropower.* Pergamon Press, Oxford, England.

CHAPTER 12

DESIGN CONSIDERATIONS FOR TIDAL SLUICE GATES FOR DRAINAGE AND FISH FARMS IN AQUACULTURE

12.1 INTRODUCTION

To alleviate drainage congestion on coastal and deltaic areas sluice gates are provided at the end of drainage channels. Tides normally enter freely and propagate upstream. During the rainy season when the channel is fed at its landward end with run-off from the drainage area, there is an interplay of fresh and tidal water. During the dry months of the year the tidal flow dominates. Significant run-off from drainage is generated only for a few days in a year. The design of drainage channels should thus be based on tidal flow modified to accommodate the upland discharge during rains. The design of a drainage channel with a sluice gate at the outfall is presented in this chapter.

Nowadays coastal areas are also utilised for aquacultural operation, for which the entry and exit of water in and out of fish farms are regulated by construction of sluice gates. So the hydraulic design for a sluice gate for aquacultural operation is also presented here. A program for computer-aided determination of sluice width for water supply to tide-fed farms is appended.

12.2 EMPIRICAL DESIGN

The discharge through the sluice at the outfall can be established from the following relationship once the period of tidal blockage is determined.

$$Q = 27 \ MI \left(\frac{24}{24 - P} \right) \qquad \qquad \dots \ (12.1)$$

where Q = discharge through the sluice in cfs; M = drainage area in sq. miles; I = drainage index in inch/day; P = period of tidal blockage in 24 hrs. As for example, let $M = 94$, $I = 3/4$ in/day, and $P = 12$ hrs 44 min. then:

$$Q = 27 \times 94 \times 3/4 \left(\frac{24}{24 - 12.7} \right) = 4050 \text{ cfs} \quad \dots (12.2)$$

Knowing the discharge the vent area can be calculated based on the following velocity relationship for flow in the barrel

$$V = 0.8 \sqrt{2gh} \quad \dots (12.3)$$

where h = depth of flow in the sluice barrel,

assume $h = 1.0$ ft, then $V = 6.43$ fps, so ventage area = 4,050/6.43 = 630 sft.

12.3 ANALYTICAL DESIGN

A drainage channel with a sluice gate at the outfall where the tides are to play has to bear some relationship with the dimension of the sluice. Accordingly neither the channel nor the sluice gate can be designed independent of each other. The sluice with rectangular vents are generally considered as broad crested weir and the following types of condition may appear, while water flows over the weir.

(i) Free flow condition and (ii) submerged flow condition when the downstream depth over the weir is less than 2/3rd of the upstream depth over the weir then the downstream water depth has no effect on the discharge over the weir. This condition is known as free flow condition. With reference to Fig. 12.1 when $H_d < 2/3\ H_u$, the free flow condition exists. The discharge at this condition is given by

$$Q = 1.72C_wB \left(H_u + \frac{V_u^2}{2g} \right)^{3/2} \quad \dots (12.4)$$

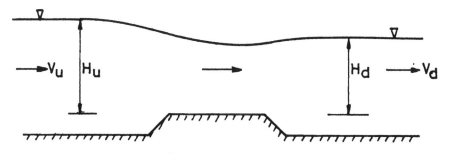

Fig. 12.1   Flow over sill

Taking the coefficient $C_w = 0.94$ and neglecting $V_u^2/2g$ being small compared to H_u, the equation becomes

$$Q = 1.6BH_u^{3/2} \qquad\qquad .... (12.5)$$

where H_u = upstream depth of sluice; B = width of openings; V_u = approach velocity.

When the downstream depth over the weir is greater than 2/3rd of the upstream depth, the condition is known as submerged condition and the performance of the opening is no longer affected by downstream depth H_d. The discharge over the weir under this condition is given by

$$Q = 4.19BH_0(H_u - H_d)^{1/2} \qquad\qquad .... (12.6)$$

neglecting the contribution of approach and exit velocities, where H_0 = sluice opening. At a condition when downstream depth is equal to the 2/3rd of upstream depth, any one of the above equations may be used for both the equations give same value.

12.4 COMPUTATION OF DISCHARGES THROUGH THE SLUICE

With the aid of the discharge formulae, it is possible to find out the discharges through a sluice where both the upstream and downstream water levels vary. The variation of downstream water levels may be due to tidal fluctuation at the outfall and the variation of upstream water level may be due to variation of basin level on account of rainfall or outflow through the sluice. Let it be considered that the tide levels as shown in Fig. 12.2 are operative at the outfall and for the simplicity consider that this tidal

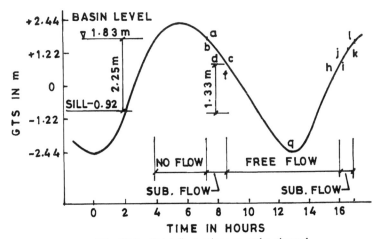

Fig. 12.2 Tidal fluctuations at outlet channel

fluctuations remain invariant due to outflow from the sluice. Let the basin water level remain invariant at +1.83 m GTS. In actual case, this level will also vary because of the imbalance of inflow and outflow from the basin. Consider the clear width of the sluice gate as B m and the sill is placed at −0.92 m GTS.

Referring to Fig. 12.2 it is seen that water from the basin cannot flow out till the tide level falls below +1.83 m GTS. As the tide level continue to fall further the free flow condition will be attained and this condition will persist till the tide level again rises to +0.92 m GTS during flood tide. The cycle will continue.

Let us now calculate the discharges through the sluice at different instants of time, a, b, c, d, e... l, as shown in Fig. 12.2. The discharges are shown in Table 12.1 below:

Table 12. 1   Discharges through the sluice at different tide levels

Time instants (see Fig. 12.2)	U/S W.L of sluice (m) GTS	D/S W.L. of sluice (m) GTS	U/S Depth of sluice (H) (m)	D/S depth of sluice (H) (m)	Type of flow	Discharge (Q) (m³/s)
a	+ 1.83	+ 1.83	2.75	2.75	No flow	0
b	+ 1.83	+ 1.53	2.75	2.44	submerged	5.69B
c	+ 1.83	+ 1.22	2.75	2.14	sumberged	6.00B
d	+ 1.83	+ 1.07	2.75	1.98	submerged	6.48B
e	+1. 83	+ 0.92	2.75	1.83	free	7.34B
f	+ 1.83	+ 0.61	2.75	1.53	free	7.34B
g	+ 1.83	− 2.44	2.75	—	free	7.34B
h	+ 1.83	+ 0.76	2.75	1.68	free	7.34B
i	+ 1.83	+ 0.92	2.75	1.83	free	7.34B
j	+ 1.83	+ 1.28	2.75	2.19	submerged	6.87B
k	+ 1.83	+ 1.53	2.75	2.44	submerged	5.69B
l	+ 1.83	+ 1.83	2.75	2.75	No flow	0

Figure 12.3 shows the discharges plotted against time. The area under the curve, a, b, c, d......k, l gives the volume of water that a sluice with a clear width B, sill level at −0.92 m GTS and constant level at +1.83 m GTS will discharge between two consecutive high water under the specified tidal variation at the downstream of the sluice.

12.5   DETERMINATION OF SILL LEVEL AND WIDTH OF SLUICE GATE

The above computations show the quantum of water that a particular sluice is capable to discharge under specified conditions. In case of design of a new channel and sluice, the sill level and the clear width are to be ascertained for discharging a specified quantity derived from the drainage area upstream of the sluice. The problem therefore reduces to: Given

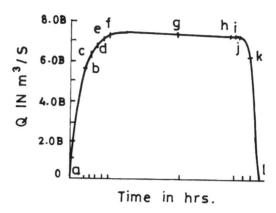

Fig. 12.3   Discharge through sluice as a function of time

(i) Basin level to be maintained, (ii) volumes of water to be drained, (iii) tide levels at the location of outflow, To find out (i) sill level at the sluice, (ii) clear width of the sluice, (iii) dimension of the drainage channel leading to the sluice. In order to do this design, it would be necessary to obtain the quantum of outflow through the sluice in terms of clear width (B) at different sill levels. Tabel 12.2 shows the volume of outflow between two consecutive high water at different sill levels and basin levels.

Figure 12.4 shows the graphical representation of the information given in Table 12.2.

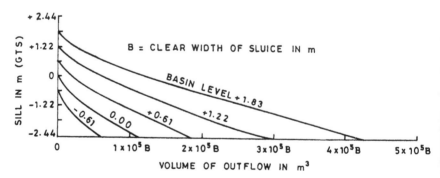

Fig. 12.4   Outflow volume as a function of still and basin level

12.6 DESIGN CONSIDERATIONS FOR FISH FARMING OPERATION

12.6.1   General

Fish farming in coastal areas also designated as brackish water aquaculture depends mostly getting its water supply from tidal creeks. The range of

Table 12.2 Outflow through the sluice at different sill and basin levels

Basin level in (m) GTS	Sill level in (m) GTS	H in (m)	Maximum discharge in m^3/s	Volume of outflow in (m^3)
+ 1.83	− 2.44	4.27	14.2B	0.43 × 10⁶ B
+ 1.83	− 1.83	3.66	11.27B	0.35 × 10⁶ B
+ 1.83	− 1.22	3.05	8.58B	0.27 × 10⁶ B
+ 1.83	− 0.92	2.75	7 .34B	0.23 × 10⁶ B
+ 1.83	− 0.61	2.44	6.14B	0.198 × 10⁶ B
+ 1.83	0.00	1.83	3.99B	0.130 × 10⁶ B
+ 1.83	0.61	1.22	2.17B	0.078 × 10⁶ B
+ 1.83	+ 1.22	0.61	0.77B	0.025 × 10⁶ B
+ 1.83	+ 1.83	0.03	0.00	0.00
+ 1.22	− 2.44	3.66	11.27B	0.288 × 10⁶ B
+ 1.22	− 1.83	3.05	8.58B	0.223 × 10⁶ B
+ 1.22	− 1.22	2.44	6.14B	0.167 × 10⁶ B
+ 1.22	− 0.61	1.83	3.99B	0.112 × 10⁶ B
+ 1.22	+ 0.00	1.22	2.17B	0.061 × 10⁶ B
+ 1.22	+ 0.61	0.61	0.77B	0.021 × 10⁶ B
+ 1.22	+ 1.22	0.00	0.00	0.00
+ .61	− 2.44	3.05	8.58B	0.187 × 10⁶ B
+ .61	− 1.83	2.44	6.14B	0.139 × 10⁶ B
+ .61	− 1.22	1.83	3.99B	0.094 × 10⁶ B
+ .61	− 0.61	1.22	2.17B	0.051 × 10⁶ B
+ .61	0.00	0.61	0.77B	0.018 × 10⁶ B
+ .61	+ .61	0.00	0.00	0.00
0.00	− 2.44	2.44	6.14B	0.111 × 10⁶ B
+ 0.00	− 1.83	1.83	3.99B	0.074 × 10⁶ B
+ 0.00	− 1.22	1.22	2.17B	0.042 × 10⁶ B
+ 0.00	− 0.61	0.61	0.77B	0.014 × 10⁶ B
+ 0.00	0.00	0.00	0.00	0.00
− 0.61	− 2.44	1.83	3.99B	0.056 × 10⁶ B
− 0.61	− 1.83	1.22	2.17B	0.033 × 10⁶ B
− 0.61	− 1.22	0.61	0.77B	0.011 × 10⁶ B
− 0.61	+ 0.61	0.00	0.00	0.00

salinity available is of the order of 18 to 25ppt and the value of pH ranges from 6.5 to 8.5. In the gravity flow system or tide fed system the supply of water to the fish farm is made by means of flood tide flow from the creeks. This system is adopted where the tides are of higher range and elevation of the foundation is sufficiently optimum to have assured supply of water into the pond during the flood flow and proper draining during the ebb flow.

Design of sluice gates for tide fed farming system have to be done considering the elevation of farm bed level, hydraulics of flow through the sluice and the tidal characteristics. Design involves the determination of sluice width for filling up the entire farm to the required depth within a specified time i.e. the specified number of tidal cycles. The procedure involves repeated calculations for various flow condition and sluice widths to achieve a satisfactory design. For this purpose a computer-aided design approach is required. For a particular site the elevation and tide characteristics remain constant. The flow conditions on the other hand will be constantly varying in accordance with the tide and pond water level. The design computations are to be carried out in steps with suitable time interval taking into account the various flow conditions with varying sluice width dimensions assumed in sequence. The sluice width which satisfies the design criterion of filling up the farm within specified tidal cycles has to be taken as the design width.

12.6.2 Theoretical Background

The elevation of pond-bed level is decided by the elevation of ground level and depth of excavation. The depth of excavation is determined by the depth of water required for the species to be cultured in that farm, the design tide and the earthwork required for the construction of dykes to the required height. The height of dyke is determined by the maximum high spring water level, rainfall in the locality etc.

For design the type of tide, time period and the tide curve in the creek at the inlet position of sluice gate is required. In the case of Indian coast the type is basically a semi-diurnal type with period equal to 12.25 hrs. The equation of the tide curve can be expressed as:

$$H = B + A \sin (Kt + \phi) \qquad \qquad \dots (12.7)$$

where, H = the tide height at any time t with respect to farm-bed level; B = distance of mean tide level from the farm bed, i.e. $(A-D)$, D being the elevation of farm bed with reference to datum, A = amplitude of the tide; K = $2\pi/T$, ϕ = phase difference = $(h - B/A)$, h being the height of water in the farm initially considered as zero.

The tidal height in the creek with reference to the farm-bed level can be expressed is shown in Fig. 12.5.

Supply of water to the farm is started by opening the sluice gate when the tidal water level in the creek is sufficiently above the farm water level generally taken as zero at the start. Various types of flow conditions will be met with through the sluice gate such as subcritical, critical and supercritical due to the continuous variation of water level on both sides of the sluice gate with time. The equations corresponding to various flow states are respectively:

$$Q = C_0 2/3 \sqrt{2 \frac{g}{3}} \left\{ \left(1 - C \frac{h^2}{H^2} \right) H \right\}^{3/2} W \qquad \dots (12.8)$$

For supercritical flow, when $h < 2/3\ H$

$$Q = C_1 2/3 \sqrt{2 \frac{g}{3}} \{3(H-h)\}^{3/2} W + C_2 (3h-2H) \sqrt{2g(H-h)}\ W \dots (12.9)$$

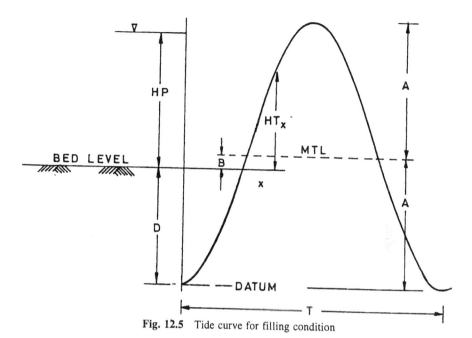

Fig. 12.5 Tide curve for filling condition

for subcritical flow, when $h > 2/3\ H$, C_0, C, C_1 and C_2 have values, 0.9761, 0.1359, 0.8890 and 0.7778 as per Kato [1]; where, h = water level in pond, H = total *head*, W = width of gates.

12.7 DESIGN APPROACH

For the design of sluice width for water supply into the tide fed farms a tentative sluice width is initially assumed. Computations are then made according to the prevailing condition after due considerations of the theoretical relation governing the flow, to find out whether the farm has been filled up to the required water level during specified tidal cycles. If the calculation is satisfactory then the adopted width is alright otherwise a new width is assumed and the whole set of operation is repeated. Since a number of trials are required the calculation, procedures are cumbersome along with attendant risk of error in computation. To obviate this an algorithm can be developed and the corresponding software will enable easy computational facilities. A typical program is placed in the appendix. For running the program the input data is required and they are as follows:

Tidal amplitude which is determined from the past data and the tidal period is obtained from the type of tide. The other inputs are area of the farm it can be assumed as fixed size. The elevation of the pond-bed level is assumed fixed by the datum and mean tide level. The initial water level in the pond is to be given as input and the final water level depends on the species to be cultured and is controlled by the available tidal amplitude.

The output of the programme are respectively, time of operation of gate in each cycle, effective head available in each tidal amplitude, number of cycles required for a particular sluice widths to fill up to required depths, total time of operation to fill upto the required depth and the values of discharge, height of tide, pond height at intervals of 30 minutes to facilitate graphical analysis.

Case study

A sample design for a tide fed farm of area 10 ha, whose elevation above the datum as 0.9 m and time period of the tide as 12.25 hrs has been made both for constant and varying tidal amplitudes. The amplitude of the tide has been taken as 1.0 m and is kept constant for all the cycles. Initial water level in the pond has been taken as zero and the final depth as 1.0 m. The program is then run for various sluice widths such as 1.0, 1.10, 1.20 m etc.

In the present case keeping the design parameter same 1.15 m width is selected as design width so that the maximum height of water required to be maintained in the farm is reached at time when the outer tide water level begins to fall and becomes equal to the inner height in other words when the inflow stops.

For the determination of sluice widths for varying amplitude it is assumed that the amplitude varies in the consecutive tidal cycles as 1.0, 1.1, and 1.2 m respectively and the design has been carried out as per other condition unchanged. In this case the design width has been found as 1 m (Fig. 12.6).

The adequacy of sluice opening for draining the farm during a period of heavy rainfall also has to be checked. For this purpose, it will be necessary to examine the maximum rainfall intensity on the farm and the outside tidal condition. Usually all available rainfall data for as large a period as available near the site have to be analysed. From the period maximum 24 hr rainfall intensities for example for the monsoon period June to September have to be ascertained. Assuming the maximum water depth to be maintained in the pond is 1.20 m from the culture point of view and the maximum rainfall depth is 25 cm during a 24 hr period, the depth of water in the pond to be drained is 1.2 + 0.25 = 1.45 m. Thereafter one has to refer to the expected high water above pond water elevation when tidal blockage will occur during monsoon months. In determining the highest tidal high water above the reference datum, one can find the duration of tidal blockage. The dimension of the sluice is determined by considering the remaining period of tidal cycle to see if the sluice is able to drain the water.

Numerical computations for the determination of required sluice size to satisfy appropriate filling and draining criteria of a brackish water fish farm

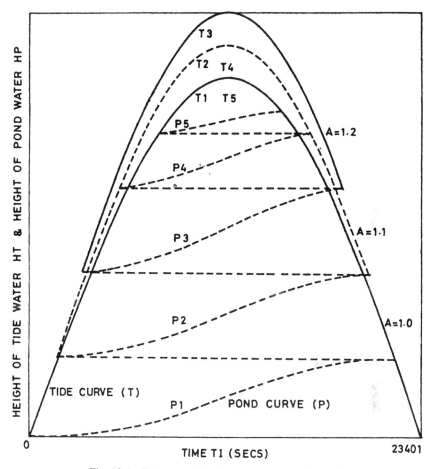

Fig. 12.6 Filling condition for varying amplitude of tide

are quite tedious as has been mentioned earlier. Programme for filling considering various hydraulic flow conditions and appropriate coefficents has already been furnished earlier. Similar programmes for draining can also be developed. Typical result showing the pond draining for a farm is shown in Figs. 12.7 and 12.8.

12.8 DETERMINATION OF SIZE OF SUPPLY-DRAINAGE CHANNEL

Brackish water from the outer sea or tidal creek is drawn into the fish pond at a specified rate and time through a channel and discharged into the sea or creek through the same canal. The channel should be designed to carry the required flow within the provided cross-section efficiently without siltation or

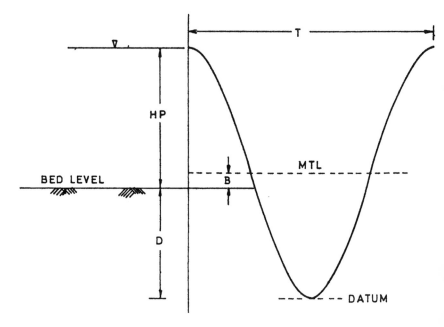

Fig. 12.7 Tide curve for draining condition

scour. The canals in fish farms are usually made of alluvial soil. They are designated as the main supply canal, secondary supply canal, drainage canal and diversion canal.

The main canal starts from the main sluice gate and usually runs through the central portion of the fish farm. In designing the size of the main canal consideration should be given to the emergency discharge of the entire farm water from the surrounding area during a period of heavy rainfall. A secondary canal is the portion between the main canal and the sluice gate of the fish pond. It is usually constructed in large farm areas. A supply channel is used as a drainage channel in many designs. However, a separate drainage channel may be provided in intensive shrimp culture which if provided is usually located at the other end of the pond, opposite and parallel to the supply canal.

The purpose of a diversion canal is to protect a farm being flooded by run-off from its watershed. The capacity of the diversion canal should be able to drain the peak run-off from the contributing watershed for a 10 to 15 year frequency storm. Its gradient should be such that the water flows towards the drainage area or around the fish farm to a convenient outlet. The main channel discharge shall be maximum required discharge passing through the sluice opening to fill up the farm area. The discharge can be

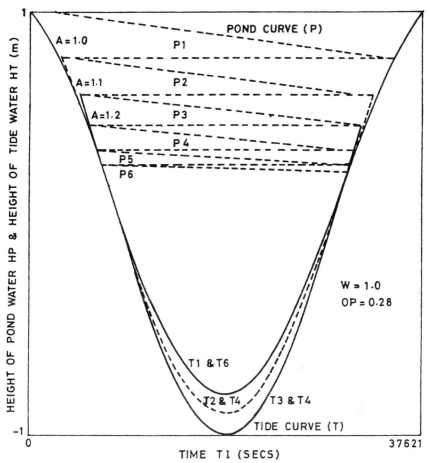

Fig. 12.8 Draining condition for varying amplitude

determined from the computations of sluice openings. Similar computations for the secondary sluice will determine its required design discharge. The capacity of supply cum drainage channels can also be obtained from considerations of the influx of tide into the pond or the farm as a tidal prism. For an individual pond the tidal prism is considered the total quantity of water rising above the pond level within the pond during the tide. For the farm as a whole the total quantity of water rising in all the ponds and channels above the pond water level should be considered.

LITERATURE CITED

1. Kato, Juichi (1980). Guide to design and construction of coastal aquacultural pond Japan International Co-operation Agency, Tokyo.
2. Kulasekaran, R. (1985). Computer-aided design of sluices for water exchange in tide farms, M. Tech, thesis, IIT, Kharagpur.
3. Ghosh, S.N. (1997). Flood Control and Drainage Engineering, Oxford and IBH Publishing Company, New Delhi.

APPENDIX: PROGRAM FOR DETERMINATION
OF SLUICE WIDTH

```
          PROGRAM FILL
          CHARACTER * 3 MODE
          OPEN (80,FILE = 'KULA1. OUT')
          WRITE (*,*)'GIVE A'
          READ (*,*) A
          WRITE (*,*) 'GIVE W'
          READ (*,*) W
          WRITE (*,*) 'GIVE T'
          READ (*,*) T
          WRITE (*,*) 'GIVE HP'
          READ (*,*) HP
          WRITE (*,*) 'GIVE AP'
          READ (*,*) AP
          WRITE (*,*) 'GIVE D'
          READ (*,*) D
          WRITE (*,*) 'GIVE HPF'
          READ (*,*) HPF
          WRITE (*,*) 'INPUT OVER'
          HP1 = 0.0
          TT0 = 0.0
          NC = 0
          I = 0
          ST = 0.0
          Y = 4.0*ATAN (1.0)
          G = 9.81
          TP = T* 3600
          B = A-D
102       TPRNT = 1.0
          NNC = 1
          IPL = 1
          IF (I.EQ,0) GO TO 103
          HP1 = HP
          WRITE (*,79)
          READ (*,44) MODE
          IF (MODE. EQ. `Y') THEN
          WRITE (*,*) `GIVE A'
          READ (*,*) A
          WRITE (*,*) `CHANGE ACCEPTED'
          ELSE
          WRITE (*,*) `OLD VALUE A =`,A, `RETAINED'
          ENDIF
103       IF (HPF. NE. 0.0) THEN
          HM = HPF
          ELSE
```

Printed and bound by CPI Group (UK) Ltd, Croydon, CR0 4YY

23/10/2024

01777667-0007